NUREG-1609

Standard Review Plan
for Transportation Packages for Radioactive Material

Manuscript Completed: March 31, 1999

Spent Fuel Project Office
Office of Nuclear Material Safety and Safeguards
U.S. Nuclear Regulatory Commission
Washington, DC 20555-0001

ABSTRACT

The Standard Review Plan for Transportation Packages for Radioactive Material provides guidance for the review and approval of applications for packages used to transport radioactive material (other than irradiated nuclear fuel) under 10 CFR Part 71.

This document is intended for use by the U.S. Nuclear Regulatory Commission (NRC) staff. Its objectives are to (1) summarize 10 CFR Part 71 requirements for package approval, (2) describe the procedures by which the NRC staff determines that these requirements have been satisfied, and (3) document the practices developed by the staff in previous reviews of package applications.

Comments, errors or omissions, and suggestions for improvement should be sent to the Director, Spent Fuel Project Office, U.S. Nuclear Regulatory Commission, Washington, DC 20555-0001.

CONTENTS

TABLES

FIGURES

ABBREVIATIONS

ANS	American Nuclear Society
ANSI	American National Standards Institute
ASME	American Society of Mechanical Engineers
ASTM	American Society for Testing and Materials
AWS	American Welding Society
B&PV	Boiler and Pressure Vessel (ASME Code)
Bq	becquerel
CFR	U.S. Code of Federal Regulations
Ci	curie
cm	centimeter
DOT	Department of Transportation
ft.	foot
in.	inch
kPa	kilopascal
m	meter
mrem	millirem
mSv	millisievert
NRC	U.S. Nuclear Regulatory Commission
PBq	petabecquerel (10^{15} Bq)
psi	pounds force per square inch
QA	quality assurance
RG	regulatory guide
SER	Safety Evaluation Report
Sv	sievert
TBq	terabecquerel (10^{12} Bq)

DEFINITIONS

The following terms are defined for the purpose of this document. The majority of terms are taken from 10 CFR 71.4 and 49 CFR 173.403, and are repeated here for convenience.

A_1	the maximum activity of special form radioactive material permitted in a Type A package.
A_2	the maximum activity of radioactive material, other than special form, low specific activity, and surface contaminated object material, permitted in a Type A package.
Carrier	a person engaged in the transportation of passengers or property by land or water as a common, contract, or private carrier, or by civil aircraft.
Certificate of compliance	a certificate issued by the NRC approving for use, with specified limitations, a specific packaging.
Close reflection by water	immediate contact by water of sufficient thickness for maximum reflection of neutrons.
Closed transport vehicle	a transport vehicle or conveyance equipped with a securely attached exterior enclosure that during normal transportation restricts the access of unauthorized persons to the cargo space containing the Class 7 (radioactive) materials. The enclosure may be either temporary or permanent, and in the case of packaged materials may be of the "see-through" type, and must limit access from the top, sides, and bottom.
Containment system	the assembly of components of the packaging intended to retain the radioactive material during transport.
Conveyance	for transport by public highway or rail, any transport vehicle or large freight container; for transport by water, any vessel or any hold, compartment, or defined deck area of a vessel, including any transport vehicle on board the vessel; and for transport by aircraft, any aircraft.

Exclusive use	the sole use by a single consignor of a conveyance for which all initial, intermediate, and final loading and unloading are carried out in accordance with the direction of the consignor or consignee. The consignor and the carrier must ensure that any loading or unloading is performed by personnel having radiological training and resources appropriate for safe handling of the consignment. The consignor must issue specific instructions, in writing, for maintenance of exclusive use shipment controls, and include them with the shipping paper information provided to the carrier by the consignor.
Fissile material	plutonium-238, plutonium-239, plutonium-241, uranium-233, uranium-235, or any combination of these radionuclides. Unirradiated natural uranium and depleted uranium, and natural uranium or depleted uranium that has been irradiated in thermal reactors only are not included in this definition. Certain exclusions from fissile material controls are provided in 10 CFR 71.53.
Fissile material package	a fissile material packaging together with its fissile material contents.
Low specific activity (LSA) material	radioactive material with limited specific activity that satisfies the descriptions and limits specified in 10 CFR 71.4.
Maximum normal operating pressure	the maximum gauge pressure that would develop in the containment system in a period of one year under the heat condition specified in 10 CFR 71.71(c)(1), in the absence of venting, external cooling by an ancillary system, or operational controls during transport.
Normal form radioactive material	radioactive material that has not been demonstrated to qualify as "special form radioactive material."
Optimum interspersed hydrogenous moderation	the presence of hydrogenous material between packages to such an extent that the maximum nuclear reactivity results.
Package	the packaging together with its radioactive contents as presented for transport.

Packaging	the assembly of components necessary to ensure compliance with the packaging requirements of 10 CFR Part 71. It may consist of one or more receptacles, absorbent materials, spacing structures, thermal insulation, radiation shielding, and devices for cooling or absorbing mechanical shocks. The vehicle, tie-down system, and auxiliary equipment may be designated as part of the packaging.
Radiation level	the radiation dose-equivalent rate expressed in millisievert(s) per hour or mSv/h (millirem(s) per hour or mrem/h). Neutron flux densities may be converted into radiation levels according to Table 1, 49 CFR 173.403.
Radioactive contents	the radioactive material within the package containment system.
Radioactive material	any material having a specific activity greater than 70 Bq per gram (0.002 microcurie per gram).
Special form radioactive material	radioactive material that satisfies the conditions specified in 10 CFR 71.4.
Specific activity of a radionuclide	the radioactivity of the radionuclide per unit mass of that nuclide. The specific activity of a material in which the radionuclide is essentially uniformly distributed is the radioactivity per unit mass of the material.
Surface contaminated object (SCO)	a solid object that is not itself classed as radioactive material, but which has radioactive material distributed on any of its surfaces. SCO must be in one of two groups with surface activity not exceeding the limits specified in 10 CFR 71.4.

Transport index	the dimensionless number (rounded up to the next tenth) placed on the label of a package, to designate the degree of control to be exercised by the carrier during transportation. The transport index is determined as follows: (1) for non-fissile material packages, the number determined by multiplying the maximum radiation level in millisievert (mSv) per hour at one meter (3.3 ft) from the external surface of the package by 100 [equivalent to the maximum radiation level in millirem per hour at one meter (3.3 ft)]; or (2) for fissile material packages, the number determined by multiplying the maximum radiation level in millisievert per hour at one meter (3.3 ft) from the external surface of the package by 100 [equivalent to the maximum radiation level in millirem per hour at one meter (3.3 ft)], or, for criticality control purposes, the number obtained as described in 10 CFR 71.59, whichever is larger.
Type A quantity	a quantity of radioactive material, the aggregate radioactivity of which does not exceed A_1 for special form radioactive material, or A_2 for normal form radioactive material, where A_1 and A_2 are given in Table A-1 of 10 CFR Part 71, or may be determined by procedures described in Appendix A of 10 CFR Part 71.
Type A packaging	a packaging approved to transport a Type A quantity of radioactive contents.
Type B packaging	a packaging approved to transport a Type B quantity of radioactive contents.
Type B quantity	a quantity of radioactive material greater than a Type A quantity.

INTRODUCTION

This document provides guidance for the review and approval of applications for packages used to transport radioactive material (other than spent fuel) under Title 10, Code of Federal Regulations, Chapter I, Part 71.

The review plan is intended primarily for use by the U.S. Nuclear Regulatory Commission (NRC) staff. Three major objectives are to:

- Summarize the regulatory requirements for package approval
- Describe the procedure by which the staff determines that the requirements have been satisfied
- Document the practices developed by the NRC in previous package certifications.

The review plan complements Regulatory Guide (RG) 7.9, revisions 1 and 2, which provide guidance to applicants on the standard format and content of applications for package approval.

This review plan does not provide an interpretation of NRC regulations within the meaning of 10 CFR 71.2. Nothing contained in this plan may be construed as having the force and effect of NRC regulations, as relieving any licensee or certificate holder from the requirements of 10 CFR Part 71 or other pertinent regulations, or as indicating that applications reviewed in accordance with this plan will necessarily be approved.

Because of the large variety of packages and the many different approaches that can be taken to evaluate these package designs, no single review plan can address in detail every situation that might be applicable to a review. The staff may therefore need to modify or expand the guidance in this review plan to adapt to specific package designs. The following areas of 10 CFR Part 71 *are not* within the scope of this review plan:

- Shipment of irradiated nuclear fuel
- Shipment of plutonium by air
- Qualification and shipment of low specific activity material and surface contaminated objects
- Qualification of special form radioactive material
- Approval of a quality assurance program
- Reports, notifications, violations, and criminal penalties
- Exemptions and general licenses
- Requirements incorporated into 10 CFR Part 71 by reference to other regulations, e.g., 10 CFR Parts 20, 21, 30, 40, 70, 73, and DOT or U.S. Postal Service regulations.

The review plan is organized at the section level in a format similar to that recommended in RG 7.9 for an application. At the subsection level, the review plan addresses the technical and regulatory bases for the review, the manner in which the review is accomplished, and findings that are generally applicable for a package that meets the approval standards. Each section follows the format below:

Subsection 1. Review Objective

This subsection states the objective of the review for each section.

Subsection 2. Areas of Review

This subsection identifies the principal areas that are reviewed to demonstrate that the package design complies with regulatory requirements. In general, the areas of review correspond to the major subsections of RG 7.9.

Subsection 3. Regulatory Requirements

This subsection summarizes the applicable regulatory requirements of 10 CFR Part 71. In many instances, the wording from the regulation is shortened, and two or more related requirements are sometimes combined for brevity. As discussed above, however, the modification in wording is not intended to change or interpret the regulations.

Subsection 4. Acceptance Criteria

This subsection includes the regulatory requirements by reference and identifies other criteria to demonstrate that the package meets the regulatory requirements.

Subsection 5. Review Procedures

This subsection provides guiding procedures for the review of a package. The review is organized in parallel with the areas of review identified in Subsection 2 above. Because of the large number of different package designs, the staff may need to expand or modify these procedures to adapt to a specific package or to address the method of evaluation presented in the application.

No section of an application for package approval is reviewed independently from information presented in other sections. For example, the Criticality Evaluation depends in part on (1) the packaging and contents description presented in the General Information section and (2) the condition of the package under the hypothetical accident condition tests in the Structural and Thermal Evaluations. Likewise, the results of the Criticality Evaluation may result in the need to implement specific Operating Procedures or Acceptance Tests. Each Review Procedures subsection of the review plan presents a schematic representation of this interface. These representations are intended only as examples; specific interfaces may vary for a particular package design.

The results of the staff review are documented in a Safety Evaluation Report which summarizes the:

- Applicable regulatory requirements

- Methods by which the application demonstrated that these requirements were met
- Staff's review of the evaluation presented in the application.

Subsection 6. Evaluation Findings

This subsection presents an example of the major finding that may be included in the Safety Evaluation Report. The staff will modify the wording as appropriate to address specific details of the application and methods of review.

Subsection 7. References

This subsection identifies references cited in the section.

Appendices

The appendices to this review plan provide detailed information on several types of packages commonly reviewed by the NRC staff. These appendices are intended to supplement information in the review plan by identifying key safety features and principal areas of review that are typical for each package type.

U.S. Code of Federal Regulations, "Packaging and Transportation of Radioactive Material," Part 71, Chapter I, Title 10, "Energy."

U.S. Nuclear Regulatory Commission, "Standard Format and Content of Part 71 Applications for Approval of Packaging for Radioactive Material," Task FC 416-4, Division 7, Proposed Revision 2 to Regulatory Guide 7.9.

U.S. Nuclear Regulatory Commission, "Standard Format and Content of Part 71 Applications for Approval of Packaging of Type B, Large Quantity, and Fissile Radioactive Material Packages," Revision to Regulatory Guide 7.9, Rev. 1.

1 GENERAL INFORMATION REVIEW

1.1 Review Objective

The objective of this review is to verify that the package design has been described in sufficient detail to provide an adequate basis for its evaluation.

1.2 Areas of Review

The package description and engineering drawings should be reviewed. The review should include:

1.2.1 Introduction

- Purpose of Application
- Summary Information

1.2.2 Package Description

- Packaging
- Containment Boundary
- Contents
- Operational Features

1.2.3 General Requirements for All Packages

- Minimum Size
- Tamper-Indicating Feature

1.2.4 Appendix

- Drawings

1.3 Regulatory Requirements

The requirements of 10 CFR Part 71 applicable to the General Information section of the application include:

- The application must include a description of the packaging design in sufficient detail to provide an adequate basis for its evaluation. [§71.31(a)(1), §71.33(a)]

- The application must include a description of the contents in sufficient detail to provide an adequate basis for evaluation of the packaging design. [§71.31(a)(1), §71.33(b)]

- The application must reference the applicant's NRC-approved quality assurance program. [§71.31(a)(3), §71.37]

- The application must identify the established codes and standards for the package design, fabrication, assembly, testing, maintenance, and use, as applicable. [§71.31(c)]

- An application for renewal of a previously approved package must be submitted no later than 30 days prior to the expiration date of the approval to assure continued use. [§71.38]

- All changes in the conditions of package approval must be approved by the NRC. An application for modification of a previously approved package may be subject to the provisions of §71.13 and §71.31(b). [§71.107(c)]

- The smallest overall dimension of the package must not be less than 10 cm (4 in.). [§71.43(a)]

- The outside of the package must incorporate a feature that, while intact, would be evidence that the package has not been opened by unauthorized persons. [§71.43(b)]

- A package containing plutonium in excess of 0.74 TBq (20 Ci) must satisfy the special containment requirements for plutonium. [§71.63]

- A fissile material package must be assigned a transport index for nuclear criticality control to limit the number of packages in a single shipment [§71.59, §71.35(b)]

- A package with a transport index greater than 10 must be transported by exclusive-use shipment. [§71.47(b), §71.59(c)]

1.4 Acceptance Criteria

- The package must meet the regulatory requirements summarized in Section 1.3.

- The package design and operation must be described in sufficient detail to provide an adequate basis for its evaluation under 10 CFR Part 71. The design must be shown on engineering drawings that can be referenced in the certificate of compliance.

1.5 Review Procedures

The review should ensure that the General Information section describes the package design and operation in sufficient detail so that the performance of the package can be evaluated in subsequent sections of the application. Figure 1-1 illustrates typical information presented in the General Information section and its relationship to the description and evaluation of the package in subsequent sections.

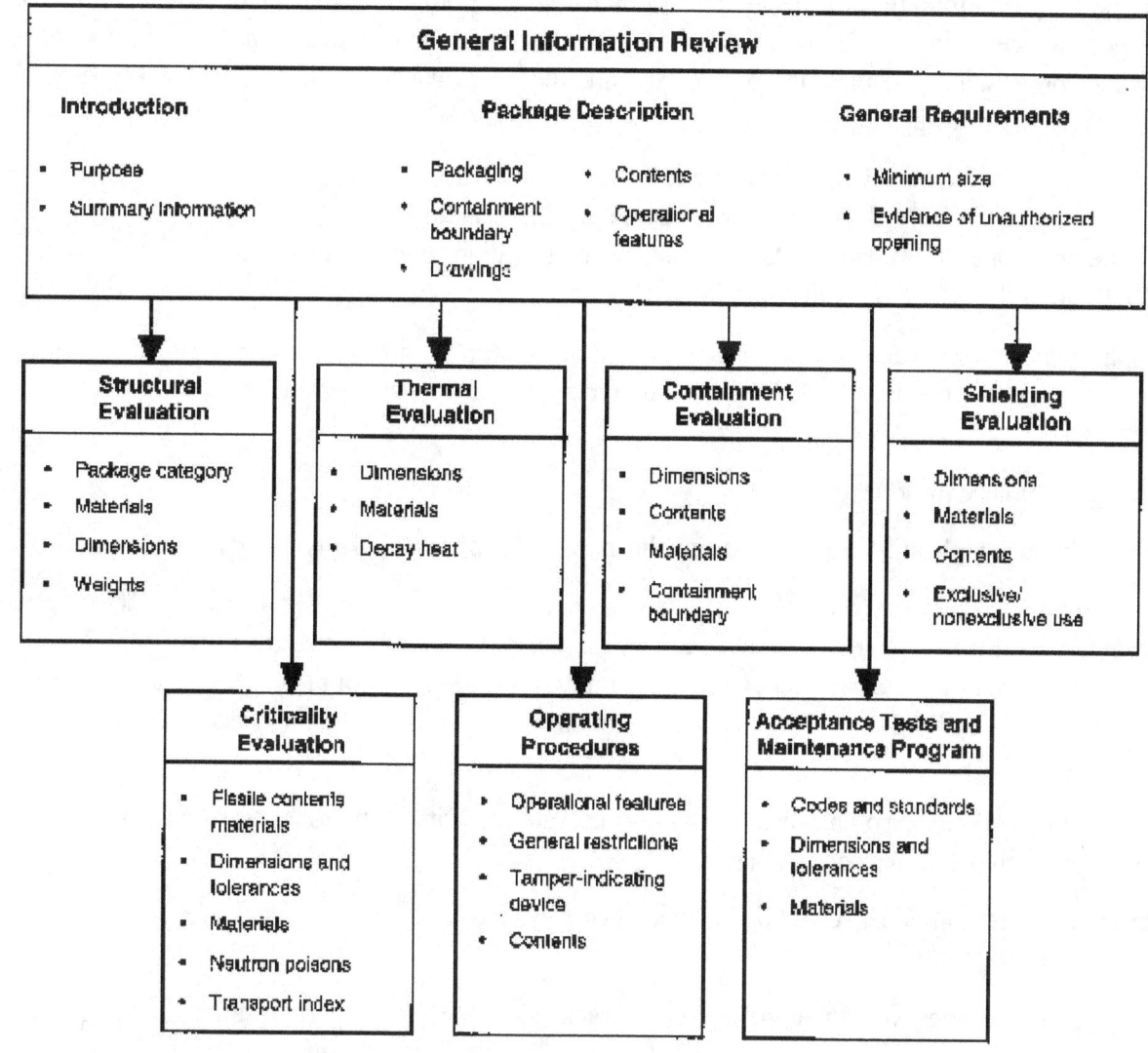

Figure 1-1 Information Flow for General Information

1.5.1 Introduction

1.5.1.1 Purpose of Application

The purpose of the application should be clearly stated. The application may be for approval of a new design, for modification of an approved design, or for renewal of an existing approval (e.g., certificate of compliance). Applications for approval of a new design should be whole and complete, and should contain the information identified in Subpart D of 10 CFR Part 71.

Applications for modification of an approved design should clearly identify the changes being requested. Modifications may include design changes, changes in authorized contents, or changes in conditions of the approval. Design changes should be clearly identified in revised packaging drawings. Packagings that do not conform to the drawings referenced in the NRC approval are not authorized for use under the general license in §71.12. Likewise, only contents specified in the approval may be transported. Package operating procedures, acceptance tests, and the maintenance program may also be specified as conditions of the approval.

Applications for modifications of an approved design should include an assessment of the requested changes and justification that these changes do not affect the ability of the package to meet the requirements of 10 CFR Part 71. Applications for modifications may be subject to the provisions of §71.13 and §71.31(b), as applicable.

Applications for renewal of an existing approval should be made prior to 30 days of expiration of the approval to assure continued use. Expiration of approvals and applications for renewal are subject to the provisions of §71.38.

1.5.1.2 Summary Information

Verify that the application references the applicant's NRC-approved quality assurance program.

Confirm that the package type and model number are designated. A new Type B package design will be designated B(U)-85 unless it has a maximum normal operating pressure greater than 700 kPa (100 psi) gauge or a pressure relief device that would allow the release of radioactive material under the tests specified in §71.73 (hypothetical accident conditions). In those cases, the package will be designated B(M)-85.

Review the intended use of the package and the maximum activity of the contents. Ensure they are consistent with the designated package type.

Ensure that any restrictions regarding the type of conveyance for shipment of the package are designated.

For Type B packages, verify that the designated package category is properly justified. Definitions of package categories are summarized in Table 1.1. Detailed justification, including calculation of an effective A_1 or A_2 from the maximum activity of the contents, might be presented in the appendix or in another section of the application (e.g., Containment). Based on the category designation, ensure that appropriate ASME code (ASME 1995) or other criteria (NUREG/CR-3019; NUREG/CR-3854), are specified for components that affect the structural integrity of containment, criticality, or shielding.

Table 1.1 Category Designations for Type B Packages (from RG 7.11)

Contents Form/ Category	Category I	Category II	Category III
Special Form	Greater than 3,000 A_1 or greater than 1.11 PBq (30,000 Ci)	Between 3,000 A_1 and 30 A_1, and not greater than 1.11 PBq (30,000 Ci)	Less than 30 A_1 and less than 1.11 PBq (30,000 Ci)

Normal Form	Greater than 3,000 A_2 or greater than 1.11 PBq (30,000 Ci)	Between 3,000 A_2 and 30 A_2, and not greater than 1.11 PBq (30,000 Ci)	Less than 30 A_2 and less than 1.11 PBq (30,000 Ci)

For fissile material packages, confirm that a transport index for criticality is designated for each contents.

Based on the transport index, determine if the transport of the package will be restricted to exclusive-use shipment.

1.5.2 Package Description

1.5.2.1 Packaging

Review the text description of the packaging. Sketches, figures, or other schematic diagrams should be provided as appropriate. Engineering drawings of the package should be presented in the appendix. Verify that the following information, as applicable, is adequately discussed:

- General packaging description, including overall dimensions, maximum weight, and minimum weight if appropriate
- Containment features (see Section 1.5.2.2)
- Neutron and gamma shielding features, including personnel barriers
- Criticality control features, including neutron poisons, moderators, and spacers
- Structural features, such as lifting and tie-down devices, impact limiters or other energy-absorbing features, internal supporting or positioning features, outer shell or outer packaging, and packaging closure devices
- Heat transfer features
- Packaging markings.

Proprietary information, such as specific design details shown on the engineering drawings, may be withheld from public disclosure subject to the provisions of 10 CFR 2.790. The request for withholding must be accompanied by an affidavit and must include information to support the claim that the material is proprietary. Requests for withholding are reviewed by the Office of the General Counsel for compliance with the requirements of 10 CFR 2.790.

1.5.2.2 Containment Boundary

Verify that the application defines the exact boundary of the containment system. This may include the containment vessel, welds, drain or fill ports, valves, seals, test ports, pressure relief devices, lids, cover plates, and other closure devices. If multiple seals are used for a single closure, the seal defined as the containment-system seal should be clearly identified. A sketch of the containment system should be provided. All components should be shown on the engineering drawings in the appendix.

If the contents include plutonium in excess of 0.74 TBq (20 Ci), the packaging must have both an inner and outer containment system unless exempted by §71.63.

Additional information regarding the review of the containment boundary is addressed in Section 4 of this review plan.

1.5.2.3 Contents

Confirm that the contents are described in the same detail as intended for the certificate of compliance. The description should include, as a minimum, the following information:

- Identification and maximum quantity (radioactivity or mass) of the radioactive material
- Identification and maximum quantity of fissile material
- Chemical and physical form, including density and moisture content, and the presence of other moderating constituents
- Location and configuration of contents within the packaging, including secondary containers, wrapping, shoring, and other material not defined as part of the packaging
- Identification and quantity of nonfissile materials used as neutron absorbers or moderators
- Any material subject to chemical, galvanic, or other reaction, including the generation of combustible gases
- Maximum normal operating pressure
- Maximum weight, and minimum weight if appropriate
- Maximum decay heat.

1.5.2.4 Operational Features

Verify that appropriate operational features are discussed. A schematic diagram of any special operational features should be included if applicable.

1.5.3 General Requirements for All Packages

Verify that the package meets the following requirements of §71.43 (General requirements for all packages):

- The smallest overall dimension of the package is not less than 10 cm (4 in.).
- The outside of a package incorporates a feature that, while intact, is evidence that the package has not been opened by unauthorized persons.

1.5.4 Appendix

Verify that information on the engineering drawings is sufficiently detailed and consistent with the package description.

Transport of radioactive materials must be authorized by license, as specified in §71.3. The general license in §71.12 authorizes licensees to transport radioactive materials in packages approved by the NRC and requires licensees to comply with the provisions of the general license, including the terms and conditions of the package approval. As noted in Section 1.5.1.1 above, packages that do not conform to the drawings specified in the NRC approval are not authorized for use.

Confirm that each drawing has a title block that identifies the preparing organization, drawing number, sheet number, title, date, and signature or initials indicating approval of the drawing. Revised drawings should identify the revision number, date, and description of the change in each revision. Proprietary information should be clearly identified. The drawings should include:

- General arrangement of packaging and contents, including dimensions
- Design features which affect the package evaluation (see Section 1.5.2.1)
- Packaging markings
- Maximum allowable weight of package
- Maximum allowable weight of contents and secondary packaging
- Minimum weights, if appropriate

Information on design features should include, as appropriate:

- Identification of the design feature and its components
- Materials of construction, including appropriate material specifications
- Codes, standards, or other similar specification documents for fabrication, assembly, and testing
- Location with respect to other package features
- Dimensions with appropriate tolerances
- Operational specifications (e.g., bolt torque)
- Weld design and inspection method.

Additional guidance on engineering drawings submitted in the application is provided in NUREG/CR-5502.

Confirm that the appendix includes a list of references and a copy of any applicable reference not generally available to the reviewer. The appendix should also provide supporting information on special fabrication procedures, determination of the package category, and other appropriate supplemental information.

1.6 Evaluation Findings

The Safety Evaluation Report does not usually include specific findings for the General Information section of the application.

1.7 References

American Society of Mechanical Engineers, *ASME Boiler and Pressure Vessel Code*, New York.

U.S. Nuclear Regulatory Commission, "Engineering Drawings for 10 CFR Part 71 Package Approvals," NUREG/CR-5502, May 1998.

U.S. Nuclear Regulatory Commission, "Fabrication Criteria for Shipping Containers," NUREG/CR-3854, March 1985.

U.S. Nuclear Regulatory Commission, "Fracture Toughness Criteria of Base Material for Ferritic Steel Shipping Cask Containment Vessels with a Maximum Wall Thickness of 4 Inches (0.1 m)," Regulatory Guide 7.11.

U.S. Nuclear Regulatory Commission, "Recommended Welding Criteria for Use in the Fabrication of Shipping Containers for Radioactive Materials," NUREG/CR-3019, March 1984.

2 STRUCTURAL REVIEW

2.1 Review Objective

The objective of this review is to verify that the structural performance of the package design has been adequately evaluated for the tests specified under normal conditions of transport and hypothetical accident conditions, and that the package has adequate structural integrity to meet the requirements of 10 CFR Part 71.

2.2 Areas of Review

The structural design of the package should be reviewed. The structural review should include the following:

2.2.1 Description of Structural Design

- Descriptive Information including Weights and Centers of Gravity
- Identification of Codes and Standards

2.2.2 Materials

- Material Properties and Specifications
- Prevention of Chemical, Galvanic, or Other Reactions
- Effects of Radiation on Materials

2.2.3 Fabrication and Examination

- Fabrication
- Examination

2.2.4 Lifting and Tie-Down Standards for All Packages

2.2.5 General Considerations

- Evaluation by Test
- Evaluation by Analysis
- Pressure

2.2.6 Normal Conditions of Transport

- Heat
- Cold
- Reduced External Pressure
- Increased External Pressure

- Vibration
- Water Spray
- Free Drop
- Corner Drop
- Compression
- Penetration

2.2.7 Hypothetical Accident Conditions

- Free Drop
- Crush
- Puncture
- Thermal
- Immersion—Fissile material
- Immersion—All packages

2.2.8 Appendix

2.3 Regulatory Requirements

Regulatory requirements of 10 CFR Part 71 applicable to the structural review are as follows:

- The package must be described and evaluated to demonstrate that it meets the structural requirements of 10 CFR Part 71. [§71.31(a)(1), §71.31(a)(2), §71.33, §71.35(a)]

- The performance of the package must be evaluated under the tests specified in §71.71 for normal conditions of transport. [§71.41(a)]

- The performance of the package must be evaluated under the tests specified in §71.73 for hypothetical accident conditions. [§71.41(a)]

- The application must identify the established codes and standards used for the structural design. [§71.31(c)]

- The package must be made of materials and construction that assure that there will be no significant chemical, galvanic or other reactions, including reactions due to possible inleakage of water, among the packaging components, among package contents, or between the packaging components and the package. The effects of radiation on the materials of construction must be considered. [§71.43(d)]

- The package must be designed, constructed, and prepared for shipment so there would be no loss or dispersal of contents, and no substantial reduction in the effectiveness of the packaging under the tests specified in §71.71 (normal conditions of transport). [§71.43(f), §71.51(a)(1)]

- The package design must meet the lifting and tie-down requirements of §71.45.

- The package design must have adequate structural integrity to meet the internal pressure test requirement specified in §71.85(b).

2.4 Acceptance Criteria

- The package must meet the regulatory requirements listed in Section 2.3.

- The package must have adequate structural integrity to meet the containment, shielding, subcriticality, and temperature requirements of 10 CFR Part 71.

2.5 Review Procedures

The structural review should ensure that the package design has been adequately described and evaluated under the normal conditions of transport and the hypothetical accident conditions to demonstrate sufficient structural integrity to meet the requirements of 10 CFR Part 71.

The structural review is based in part on the descriptions and evaluations presented in the General Information and the Thermal Evaluation sections of the application. Similarly, results of the structural review are considered in the review of all other sections of the application. An example of this information flow for the structural review is shown in Figure 2-1.

2.5.1 Description of Structural Design

2.5.1.1 Descriptive Information Including Weights and Centers of Gravity

Review the package description presented in the General Information and Structural Evaluation sections of the application. Descriptive information important to structures includes:

- Dimensions, tolerances, and materials
- Maximum and minimum weights and centers of gravity of packaging and major sub-assemblies

- Maximum and minimum weight of contents, if appropriate
- Maximum normal operating pressure
- Description of closure system
- Description of handling requirements
- Fabrication methods, as appropriate.

Confirm that the text and sketches describing the structural design features are consistent with the engineering drawings and the models used in the structural evaluation.

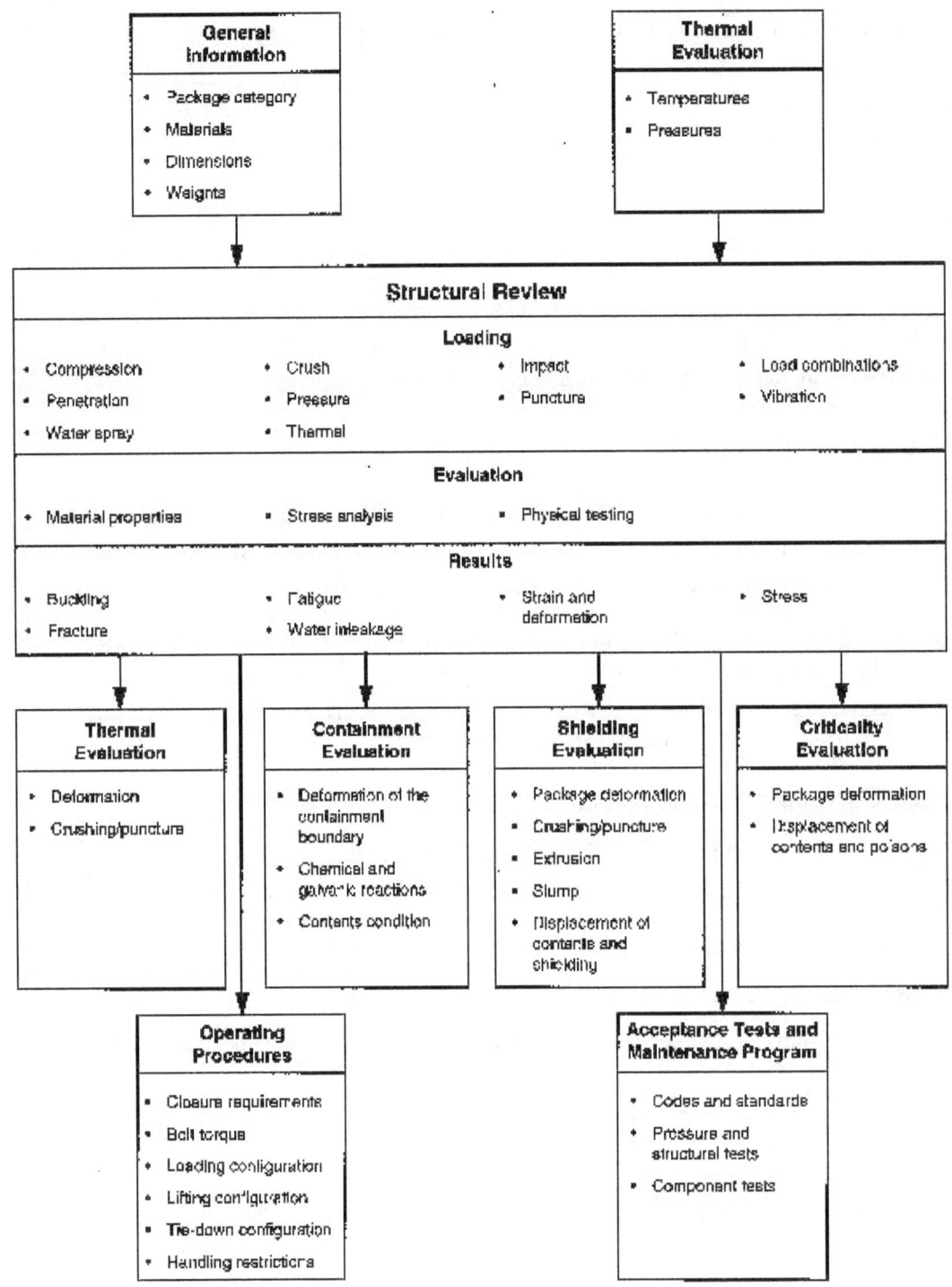

Figure 2-1 Information Flow for the Structural Review

2.5.1.2 Identification of Codes and Standards for Package Design

Review the codes and standards that have been identified by the applicant. The review should include an assessment of the applicability of the codes and standards that the applicant has identified. The assessment may include the following considerations:

- The code or standard was developed for structures or features of similar design and material, if not specifically for shipping packages.
- The code or standard was developed for structures with similar loading conditions.
- The code or standard was developed for structures that have similar consequences of failure.
- The code or standard adequately addresses potential failure modes.
- The codes or standard addresses margins of safety.

The code or standard should consider the package category (see Table 1.1) as appropriate. NUREG/CR-3854 provides a list of industrial codes and standards which may be used for fabricating components of transportation packagings according to package category.

Several regulatory guides and NUREGs provide additional design guidance for evaluation of packages using information from existing codes and practices. RG 7.8 identifies the load combinations to be used in package design evaluation. RG 7.6 provides design criteria for the containment system of Type B packages. NUREG/CR-4554, Vol. 6 discusses the buckling evaluation of containment vessels. NUREG/CR-3019 presents criteria for transportation package welding.

2.5.2 Materials

2.5.2.1 Material Properties and Specifications

Review the properties of the materials of construction. An appropriate specification should be identified for the control of each material. Materials and their properties should be consistent with the design code or standard selected. If no standard is available, the application should provide adequately documented material properties and specifications for the design and fabrication of the packaging.

Verify that the materials of structural components have sufficient fracture toughness to preclude brittle fracture under normal conditions of transport and hypothetical accident conditions. RGs 7.11 and 7.12 provide criteria for fracture toughness.

Verify that the material properties are appropriate for the load conditions (e.g., static or dynamic impact loading, hot or cold temperatures, and wet or dry conditions). Verify that appropriate temperatures at which allowable stress limits are defined are consistent with minimum and maximum service temperatures. Verify that the force-deformation properties for impact limiters are based on appropriate test conditions and temperature.

2.5.2.2 Prevention of Chemical, Galvanic, or Other Reactions

Review the materials and coatings of the package to verify that they will not produce a significant chemical or galvanic reaction among packaging components, among packaging contents, or between the packaging components and the package contents. The review should consider reactions resulting from inleakage of water. Evaluate the possible generation of hydrogen and other flammable gases. Galvanic interactions and the formation of eutectics should be considered for metallic components that may come into physical contact with one another. Such interactions may occur with depleted uranium, lead, or aluminum in contact with steel.

2.5.2.3 Effects of Radiation on Materials

Verify that any damaging effects of radiation on the packaging materials have been appropriately considered. These effects may include degradation of seals, sealing materials, coatings, adhesives, and structural materials.

2.5.3 Fabrication and Examination

2.5.3.1 Fabrication

Fabrication generally addresses:

- Forming, fitting, and aligning
- Welding and brazing
- Heat treatment
- Mechanical joints.

If fabrication specifications are prescribed by an acceptable code or standard (e.g., ASME, AWS), the code or standard should be identified on the engineering drawings (NUREG/CR-5502). Unless the application justifies otherwise, specifications of the same code or standard used for design should also be used for fabrication. For components for which no code or standard is applicable, the application should identify the specifications on which the evaluation depends and describe the method of control to assure that these specifications are achieved. This description may reference a quality assurance or other appropriate specifications document, which should be specified on the engineering drawings. As noted in Section 1 of this review plan, the engineering drawings are generally included as conditions of approval in the certificate of compliance.

2.5.3.2 Examination

Examination addresses the methods and criteria by which the fabrication is determined to be acceptable. Unless the application justifies otherwise, specifications of the same code or standard used for fabrication should also be used for examination. For components for which no fabrication code or standard is applicable, the application should summarize the examination methods and acceptance criteria in Section 8, Acceptance Tests and Maintenance Procedures. As noted in Section 8 of this review plan, acceptance tests are generally included as conditions of approval in the certificate of compliance.

2.5.4 Lifting and Tie-Down Standards for All Packages

2.5.4.1 Lifting Devices

Review the design and evaluation of lifting devices that are a structural part of the package, their connection with the package body, and the package body in the local area around the lifting devices. Verify that the evaluation demonstrates these devices comply with the requirements of §71.45(a), including failure under excessive load.

2.5.4.2 Tie-Down Devices

Review the design and evaluation of tie-down devices that are a structural part of the package, their connection with the package body, and the package body in the local area around the tie- down devices. Verify that the evaluation demonstrates these devices comply with the requirements of §71.45(b), including failure under excessive load.

2.5.5 General Considerations

The evaluation should demonstrate that the structural performance of the package meets the criteria discussed in Section 2.5.6 for normal conditions of transport and in Section 2.5.7 for hypothetical accident conditions.

- The most limiting initial conditions have been used (see RG 7.8 for guidance on initial condition selection).
- The evaluation methods are appropriate for the loading conditions considered and follow accepted practices and precepts.
- The interpretations of results are correct.
- The most damaging orientations have been considered. The most damaging orientation for one component may not be the most damaging for another component.
- Design criteria such as those provided in RG 7.6 have been applied.

2.5.5.1 Evaluation by Test

If the package is evaluated by testing, the review should include at least the following:

- Review the description of the surface (e.g., material, mass, dimensions) used for the free drop and crush test. Confirm that it represents an essentially unyielding surface as specified in §71.73(c)(1).
- Review the description of the steel bar (e.g., material, dimensions, orientation, method of mounting) used for the puncture test. Confirm that it is securely attached to an essentially unyielding surface, has sufficient length to cause maximum damage to the package, and meets the other specifications of §71.73(c)(3).

- Verify that the test specimen has been fabricated using the same materials, methods, and quality assurance as specified in the design. Any differences should be identified and the effects evaluated in the application. Substitutes for the contents should have the same representative weight as the actual contents.

- Verify that the selected drop orientations consider the orientations for which maximum damage is expected, and that the selection is justified.

- Verify that all test results are evaluated and their implications interpreted, including both interior and exterior damage of the test article. Unexpected or unexplainable test results indicating possible testing problems or non-reproducible specimen behavior should be discussed and evaluated.

- Review the video and photos of the tests, if available.

- Verify that the tests demonstrate an adequate margin of safety. The test results should clearly show that the effects of the tests can be reliably reproduced. Verify that the description of the test results includes a discussion of the effects of uncertainties in mechanical properties, test conditions, and diagnostics.

- Review the criteria for evaluating pass / fail for the test conditions. Compare the test results with these criteria.

2.5.5.2 Evaluation by Analysis

If the application provides evaluation by analysis, the review should include at least the following:

- Verify that a clear description of the calculation, and all assumptions, are included (see RG 7.6 for guidance on design criteria for analysis).

- Verify that the response of the package to loads, in terms of stress and strain to components and structural members, is shown, and that the structural stability of individual members, as applicable, is evaluated.

- Verify that the analytical method considers impact at any angle, rigid-body rotation, and secondary impact (slap down).

- Verify that the computer codes, if used, are valid for the intended application, use methods that are consistent with standard practice and procedures, and are benchmarked.

- Verify that a dynamic amplification factor has been appropriately applied if a quasi-static analysis technique has been used. A summary of quasi-static and dynamic analysis methods for impact analysis is provided in NUREG/CR-3966.

- Verify that the models and material properties are appropriate for the load combinations considered. Ensure that the material properties (e.g. elastic, plastic) are consistent with the analysis methods. The application should justify the strain rate at which the properties were determined. Confirm that the analysis considers true stress-strain or engineering stress-strain, as applicable.

- Review the summary table of the results of the analyses, compare the results with the acceptance criteria provided, and verify that the acceptance criteria have been met. Verify that the criteria are in accordance with appropriate codes and standards.

2.5.5.3 Pressure

Prior to first use of each packaging with a maximum normal operating pressure exceeding 35 kPa (5 psi) gauge, the containment system must be pressure tested at 150% of its maximum normal operating pressure in accordance with §71.85(b). Confirm that the analysis of this acceptance test is provided in the application.

2.5.6 Structural Evaluation under Normal Conditions of Transport

The evaluation of the package under the normal conditions of transport is based on the effects of the tests and conditions specified in §71.71. These tests must not result in a decrease in package effectiveness. For example, there should be:

- No loss or dispersal of contents
- No structural changes reducing the effectiveness of components required for shielding, for heat transfer, or for maintaining subcriticality or containment
- No changes to the package affecting its ability to withstand the hypothetical accident conditions.

The ambient air temperature before and after the tests must remain at that value between -29°C (-20°F) and +38°C (100°F) which is most unfavorable for the feature under consideration. The initial internal pressure in the containment vessel must be considered to be the maximum normal operating pressure, unless a lower internal pressure consistent with the selected ambient temperature is less favorable.

2.5.6.1 Heat

Confirm that the evaluation of thermal performance and the maximum temperatures under the heat condition are consistent with the Thermal Evaluation section.

The evaluations should consider the maximum normal operating pressure in combination with the maximum internal heat load and any residual fabrication stresses. Verify that any differential thermal expansions and possible geometric interferences have been considered.

If the structural design has been evaluated by engineering analysis, verify that the stresses are within the limits for normal condition loads.

2.5.6.2 Cold

Confirm that the evaluation of thermal performance and the temperatures under the cold test condition are consistent with the Thermal section.

The evaluations should consider the minimum internal pressure with the minimum internal heat load (typically assumed to be no decay heat) and any residual fabrication stresses. Verify that differential thermal expansions that could result in possible geometric interferences have been considered. Verify that possible freezing of liquids has been considered.

Verify that the stresses are within the limits for normal condition loads.

2.5.6.3 Reduced External Pressure

Determine that the application adequately evaluates the package design for the effects of reduced external pressure equal to 25 kPa (3.5 psi) absolute. Verify that the application considers the greatest possible pressure difference between the inside and outside of the package as well as the inside and outside of the containment system.

2.5.6.4 Increased External Pressure

Determine that the application adequately evaluates the package design for the effects of increased external pressure equal to 140 kPa (20 psi) absolute. Verify that the application considers this loading condition in combination with minimum internal pressure. Verify that the application considers the greatest possible pressure difference between the inside and outside of the package as well as the inside and outside of the containment system. Consider the possibility of buckling (see NUREG/CR-4554, Vol. 6).

2.5.6.5 Vibration

Determine that the application adequately evaluates the package design for the effects of vibration normally incident to transport. A fatigue analysis should be provided for highly stressed systems, considering the combined stresses due to vibration, temperature, and pressure loads. If closure bolts are reused, verify that the bolt preload is included in the fatigue evaluation. NUREG/CR-6007 provides guidance on bolt evaluation. Verify that a resonant vibration condition, which can cause rapid fatigue damage, is not present in any packaging component. The effect on package internals should be considered. Additional guidance for vibration evaluation is provided in NUREG/CR-2146 and NUREG/CR-0128.

2.5.6.6 Water Spray

Review the package design for the effects of the water spray test. Verify that this test has no significant effect on material properties.

2.5.6.7 Free Drop

Review the package design for the effects of the free drop test. The application should address factors such as drop orientation, effects of free drop in combination with pressure, heat, and cold temperatures, and other factors discussed in Section 2.5.5.

Review the evaluation of the closure lid bolt design for the combined effects of free drop impact force, internal pressures, thermal stress, O-ring compression force, and bolt preload. Bolt evaluation methods are presented in NUREG/CR-6007.

Review the evaluation of other package components, such as port covers, port cover plates, and shield enclosures, for the combined effects of package drop impact force, internal pressures, and thermal stress.

2.5.6.8 Corner Drop

Review the package design for the effects of the corner drop test, if applicable.

2.5.6.9 Compression

Review the package design for the effects of the compression test, if applicable.

2.5.6.10 Penetration

Review the evaluation of the package for the penetration test. Verify that the application considers the package location which is most vulnerable.

2.5.7 Structural Evaluation under Hypothetical Accident Conditions

The evaluation under hypothetical accident conditions must be based on sequential application of the tests specified in §71.73, in the order indicated, to determine their cumulative effect on a package. The evaluation of the ability of a package to withstand any one test must consider the damage that resulted from the previous tests. In addition, as stated in Section 2.5.6, the tests under normal conditions of transport must not affect the package's ability to withstand the hypothetical accident condition tests.

Confirm that the evaluation demonstrates that the package has adequate structural integrity to satisfy the containment, shielding, and subcriticality requirements of 10 CFR Part 71 under the hypothetical accident conditions:

- Inelastic deformation of the containment closure and seal system is generally unacceptable for the containment evaluation.
- Deformation of shielding components should be reviewed in terms of the shielding evaluation.
- Deformation of components required for heat transfer or insulation should be reviewed in terms of the thermal evaluation.
- Deformation of components required for subcriticality should be reviewed in terms of the criticality evaluation.

With respect to the initial conditions for the tests (except for the water immersion tests), the ambient air temperature before and after the tests must remain at that value between -29°C (-20°F) and +38°C (100°F) which is most unfavorable for the feature under consideration. The initial internal pressure within the containment system must be the maximum normal operating pressure, unless a lower internal pressure consistent with the selected ambient temperature is less favorable.

2.5.7.1 Free Drop

Review the evaluation of the free drop. Verify that structural integrity has been evaluated for the drop orientation which causes the most severe damage, including center-of-gravity-over-corner, oblique orientation with secondary impact (slap down), side drop, and drop onto the closure. The most damaging orientation for one component might not be the most damaging orientation for another component. If a feature such as a tie-down component is a structural part of the package, it should be considered in the selection of the drop test configurations and the drop orientation.

For a package with lead shielding, the effects of lead slump should be evaluated. The lead slump determined should be consistent with that used in the shielding evaluation. Lead slump is discussed in

NUREG/CR-4554, Vol. 3.

Review the evaluation of the closure lid bolt design for the combined effects of free drop impact force, internal pressures, thermal stress, O-ring compression force, and bolt preload. Bolt evaluation methods are presented in NUREG/CR-6007.

Review the evaluation of other package components, such as port covers, port cover plates, and shield enclosures, for the combined effects of package drop impact force, internal pressures, and thermal stress.

Buckling of package components should be considered. Evaluation of containment vessel buckling is discussed in NUREG/CR-4554, Vol. 6.

2.5.7.2 Crush

If applicable, review the evaluation of the package for the dynamic crush condition. Verify that the choice of the most unfavorable orientation has been justified.

2.5.7.3 Puncture

Review the evaluation of the package for the puncture test. Verify that the position for which maximum damage would be expected has been identified and justified. Any damage resulting from the free drop and crush conditions must be considered when evaluating this test.

Although analytical methods are available for predicting puncture, empirical formulas derived from puncture test results of laminated panels are usually used for the design of packages. The Nelm's formula developed specifically for package design provides the minimum thickness needed for preventing the puncture of the steel surface layer of a typical steel-lead-steel laminated cask wall. A description of methods for puncture evaluation is provided in NUREG/CR-4554, Vol. 7. Additional considerations in puncture testing are identified in NRC Bulletin 97-02.

Verify that punctures at oblique angles, near a support, at a valve, and at a penetration have been considered, as appropriate.

2.5.7.4 Thermal

Verify that the structural design is evaluated for the effects of a fully engulfing fire, as specified in §71.73(c)(4). Any damage resulting from the free drop, crush, and puncture conditions must be incorporated into the initial condition of the package for the fire test. Confirm that the determination of the maximum pressure in the package during or after the test considers the temperatures resulting from the fire and any increase in gas inventory caused by combustion or decomposition processes. Verify that the maximum thermal stresses, which can occur either during or after the fire, are evaluated and are consistent with the Thermal Evaluation section of the application.

2.5.7.5 Immersion—Fissile Material

If the contents include fissile material subject to the requirements of §71.55, and if water inleakage has not been assumed for the criticality analysis, review the evaluation of the test of a damaged specimen immersed under a head of water of at least 0.9 m (3 ft.) in the orientation for which maximum leakage is expected.

2.5.7.6 Immersion—All Packages

Review the evaluation of a separate, undamaged specimen subjected to water pressure equivalent to immersion under a head of water of at least 15 m (50 ft.). For test purposes, an external pressure of water of 150 kPa (21.7 psi) gauge is considered to meet these conditions.

2.5.8 Appendix

Confirm that the appendix includes a list of references, copies of applicable references if not generally available to the reviewer, computer code descriptions, input and output files, test results, and other appropriate supplemental information.

If the package is evaluated by test, review the description. The description should include:

- Test procedures
- Test package description
- Test initial and boundary conditions
- Test chronologies (planned and actual)
- Photographs of the package components, including any structural damage, before and after the tests
- Test measurements, including, at a minimum, documentation of test package physical changes as a result of the tests
- Test results
- Methods used to obtain these corrected results.

2.6 Evaluation Findings

The Safety Evaluation Report should include a finding similar to the following:

> Based on review of the statements and representations in the application, the staff concludes that the structural design has been adequately described and evaluated and that the package has adequate structural integrity to meet the requirements of 10 CFR Part 71.

2.7 References

U.S. Nuclear Regulatory Commission, "Design Criteria for the Structural Analysis of Shipping Cask Containment Vessels," Regulatory Guide 7.6.

U.S. Nuclear Regulatory Commission, "Dynamic Analysis to Establish Normal Shock and Vibration of Radioactive Material Shipping Packages, Volume 3: Final Summary Report," NUREG/CR-2146, Vol. 3, October 1983.

U.S. Nuclear Regulatory Commission, "Engineering Drawings for 10 CFR Part 71 Package Approvals," NUREG/CR-5502, May 1998.

U.S. Nuclear Regulatory Commission, "Fabrication Criteria for Shipping Containers," NUREG/CR-3854, March 1985.

U.S. Nuclear Regulatory Commission, "Fracture Toughness Criteria of Base Material for Ferritic Steel Shipping Cask Containment Vessels with a Maximum Wall Thickness of 4 Inches (0.1m)," Regulatory Guide 7.11.

U.S. Nuclear Regulatory Commission, "Fracture Toughness Criteria of Base Material for Ferritic Steel Shipping Cask Containment Vessels with a Wall Thickness Greater than 4 Inches (0.1m)," Regulatory Guide 7.12.

U.S. Nuclear Regulatory Commission, "Load Combinations for the Structural Analysis of Shipping Casks for Radioactive Material," Regulatory Guide 7.8.

U.S. Nuclear Regulatory Commission, "Methods for Impact Analysis of Shipping Containers," NUREG/CR-3966, November 1987.

U.S. Nuclear Regulatory Commission, "Puncture Testing of Shipping Packages under 10 CFR Part 71," Bulletin 97-02, September 23, 1997.

U.S. Nuclear Regulatory Commission, "Recommended Welding Criteria for Use in the Fabrication of Shipping Containers for Radioactive Materials," NUREG/CR-3019, March 1985.

U.S. Nuclear Regulatory Commission, "SCANS (Shipping Cask ANalysis System): A Microcomputer Based Analysis System for Shipping Cask Design Review," NUREG/CR-4554, February 1990.

U.S. Nuclear Regulatory Commission, "Shock and Vibration Environments for a Large Shipping Container During Truck Transport (Part II)," NUREG/CR-0128, August 1978.

U.S. Nuclear Regulatory Commission, "Stress Analysis of Closure Bolts for Shipping Casks," NUREG/CR-6007, January 1993.

3 THERMAL REVIEW

3.1 Review Objective

The objective of this review is to verify that the thermal performance of the package design has been adequately evaluated for the thermal tests specified under normal conditions of transport and hypothetical accident conditions, and that the package design meets the thermal performance requirements of 10 CFR Part 71.

3.2 Areas of Review

The description and evaluation of the package thermal design should be reviewed. The thermal review should include the following:

3.2.1 Description of Thermal Design

- Design Features
- Contents Decay Heat
- Summary Tables of Temperatures
- Summary Tables of Maximum Pressures in the Containment System

3.2.2 Material Properties and Component Specifications

- Material Thermal Properties
- Component Specifications

3.2.3 General Considerations

- Evaluation by Analysis
- Evaluation by Test
- Margins of Safety

3.2.4 Thermal Evaluation under Normal Conditions of Transport

- Heat and Cold
- Maximum Normal Operating Pressure
- Maximum Thermal Stresses

3.2.5 Thermal Evaluation under Hypothetical Accident Conditions

- Initial Conditions
- Fire Test Conditions
- Maximum Temperatures and Pressure
- Maximum Thermal Stresses

3.2.6 Appendix

- Description of Test Facilities
- Test Results
- Applicable Supporting Documents or Specifications
- Analyses Details

3.3 Regulatory Requirements

Regulatory requirements of 10 CFR Part 71 applicable to the thermal evaluation are as follows:

- The package design must be described and evaluated to demonstrate that it satisfies the thermal requirements of 10 CFR Part 71. [§71.31(a)(1), §71.31(a)(2), §71.33, §71.35(a)]

- The application must identify the established codes and standards used for the thermal design. [§71.31(c)]

- The performance of the package must be evaluated under the tests specified in §71.71 for normal conditions of transport and §71.73 for hypothetical accident conditions. [§71.41(a)]

- The package must be designed, constructed, and prepared for transport so that there will be no significant decrease in packaging effectiveness under the tests specified in §71.71 (normal conditions of transport). [§71.43(f), §71.51(a)(1)]

- The package must be designed, constructed, and prepared for transport so that the accessible surface temperature does not exceed the regulatory limits. [§71.43(g)]

- The package design must not rely on mechanical cooling systems to meet containment requirements. [§71.51(c)]

3.4 Acceptance Criteria

- The package design must meet the regulatory requirements listed in Section 3.3.

- The package must have adequate thermal performance to meet the containment, shielding, subcriticality, and temperature requirements of 10 CFR Part 71, under normal conditions of transport and hypothetical accident conditions.

3.5 Review Procedures

The thermal review should ensure that the package design has been described and evaluated for the thermal tests specified under normal conditions of transport and hypothetical accident conditions, and that it meets the thermal performance requirements of 10 CFR Part 71.

The thermal review is based in part on the descriptions and evaluations presented in the General Information and Structural Evaluation sections of the application. Similarly, results of the thermal review are considered in the review of several other sections of the application. An example of information flow for the thermal review is shown in Figure 3-1.

3.5.1 Description of Thermal Design

3.5.1.1 Design Features

Review the thermal design features presented in the General Information and Thermal Evaluations sections of the application. Design features important to thermal performance include:

- Package geometry and materials of construction
- The structural and mechanical features that may affect heat transfer, such as cooling fins, insulating materials, surface conditions of the package components, and gaps or physical contacts between internal components.

3.5.1.2 Contents Decay Heat

Verify that the maximum decay heat and the radioactivity of the contents are consistent with those in the General Information section. Ensure that the decay heat is properly determined from the maximum allowed radioactive contents.

3.5.1.3 Summary Tables of Temperatures

Confirm that summary tables of the maximum or minimum temperatures that affect structural integrity, containment, shielding, and criticality are presented for both normal conditions of transport and hypothetical accident conditions. For the fire test condition, the tables should also include:

- The maximum temperatures and the time at which they occur after fire initiation
- The maximum temperatures of the post-fire steady-state condition.

Confirm that these temperatures are consistent with those of the Structural Evaluation and Containment section.

3.5.1.4 Summary Tables of Maximum Pressures in the Containment System

Verify that summary tables include the maximum normal operating pressure and the maximum pressure under hypothetical accident conditions. Confirm that these pressures are consistent with those in the General Information, Structural Evaluation, Containment, and Acceptance Tests and Maintenance Program sections.

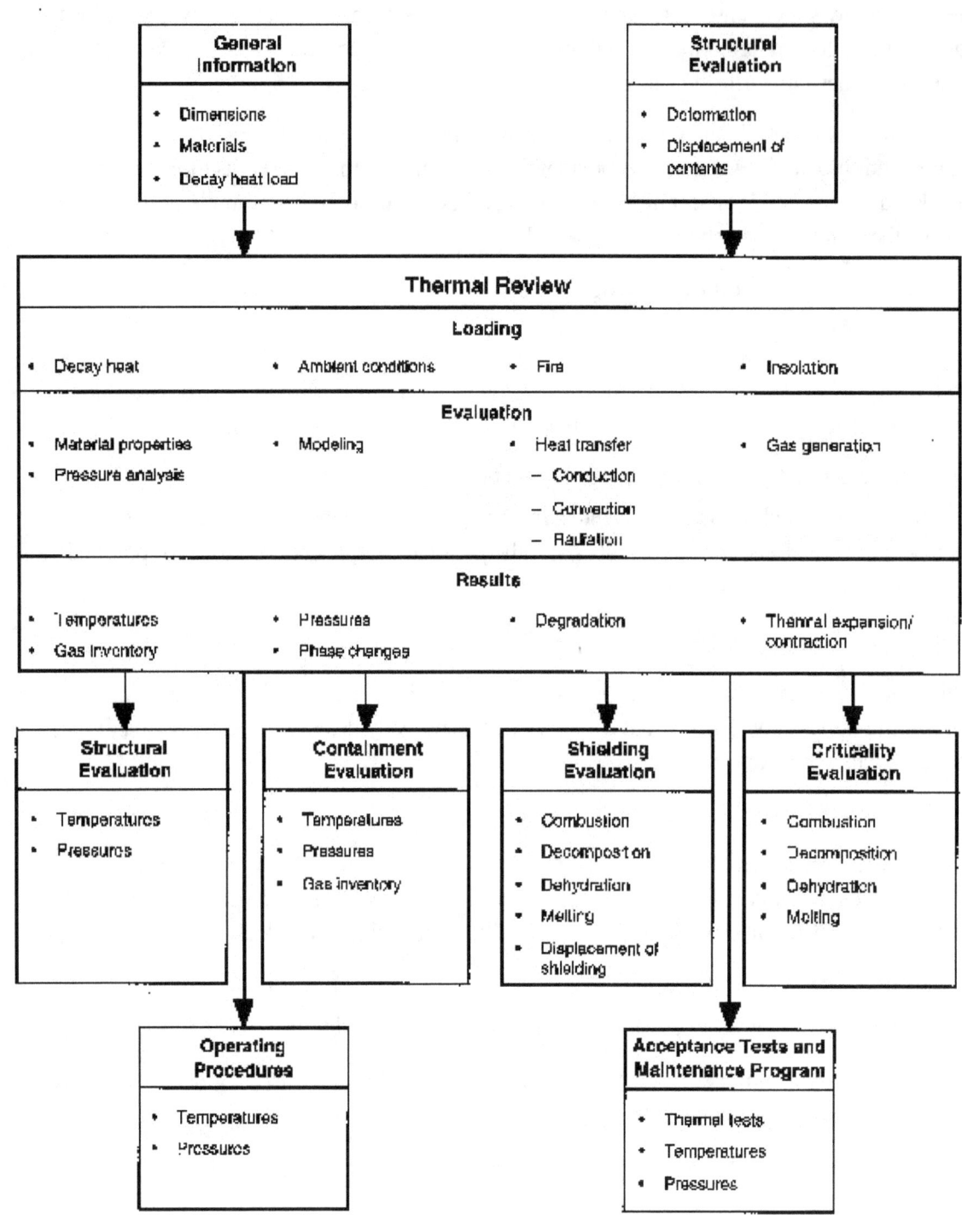

Figure 3-1 Information Flow for the Thermal Review

3.5.2 Material Properties and Component Specifications

3.5.2.1 Material Properties

Verify that the appropriate thermal properties are specified for materials that affect heat transfer both within the package and from the package to the environment. These materials include any liquids or gases within the package and gases external to the package for hypothetical accident conditions. Confirm that the thermal absorptivities and emissivities are appropriate for the package surface conditions and each thermal condition being evaluated. If a property is reported as a single value, ensure that this value bounds the equivalent temperature-dependent property.

3.5.2.2 Component Specifications

Confirm that the maximum allowable service temperatures or pressures are specified for each package component, as appropriate. Verify that the minimum allowable service temperature of all components is less than or equal to -40°C (-40°F). Ensure that technical specifications are provided for applicable package components (e.g., pressure relief valves and fusible plugs).

3.5.3 General Considerations

Thermal evaluations of the package design can be performed by either analysis or test, or by a combination of both.

3.5.3.1 Evaluation by Analysis

Confirm that the methods of thermal analysis are sufficiently described to permit review and independent verification. Ensure that for each thermal analysis:

- The methods used are properly referenced or developed in the application.
- These methods are correctly applied.
- Assumptions in modeling heat sources and heat transfer paths and modes are clearly stated and are justified.
- The appropriate thermal properties for the package materials are correctly incorporated into the thermal evaluations.
- Appropriate expressions are used for conductive, convective, and radiative heat transfer among package components and from the surfaces of the package to the environment.
- The evaluation considers changes in package geometry and material properties resulting from structural and thermal tests under normal conditions of transport and hypothetical accident conditions.
- The required temperature and thermal boundary conditions for normal conditions of transport and hypothetical accident conditions are correctly applied.

- The time interval after the fire test is adequate to assure that maximum component temperatures and post-fire steady-state temperatures have been achieved.

- The results from the thermal evaluations are clearly presented in adequate detail in a combination of figures, tables, and supporting text.

- The maximum temperature and pressure of each component do not exceed their allowable value.

- Combustion of package components is considered, including the heat produced.

The thermal evaluation should assume that the heat transfer medium is air, and the effects of air on the contents and packaging components (e.g., oxidation of depleted uranium shielding) should be addressed.

3.5.3.2 Evaluation by Test

Verify that the test package, test facility, and test procedures are described in adequate detail. Confirm that the test package was fabricated, the test facility operated, and the test results evaluated under proper quality assurance programs. Verify that the test package has been adequately designed.

- The thermal performance of the test package, including simulated package contents and any attached test instrumentation and mounting hardware, should be representative of the actual package design.

- The temperature-sensing instrumentation should be located to measure the appropriate maximum package component temperatures and characterize the significant heat transfer pathways.

- Test package instrumentation (such as temperature- or pressure-sensing devices) should be mounted at locations that minimize their effects on local test package temperatures.

Review the ability of both the test facility (pool-fire or furnace facility) and the test procedures to meet the range of thermal conditions (e.g., insolation and fire heat fluxes or temperatures). Additional guidance for review of thermal testing is presented in Section 3.5.6.

Verify that the appropriate results from normal conditions of transport and hypothetical accident condition thermal tests are adequately presented:

- Initial conditions (e.g., temperatures, pressures), and changes in the package resulting from structural tests

- Maximum steady-state temperatures or pressures (e.g., hot normal conditions of transport, pre-fire conditions)

- Maximum temperatures and pressures during the fire and post-fire periods

- Physical changes in the package condition resulting from the fire test, such as changes in package material properties caused by combustion or melting of package components.

Some conditions, such as ambient temperature, decay heat of the contents, or package emissivity or absorptivity, may not be exactly represented in a thermal test. The thermal evaluation should include appropriate corrections or evaluations to account for these differences. For example, the thermal evaluation should include a temperature correction if the ambient temperature at the onset of the fire test was lower than 38°C (100°F).

3.5.3.3 Margins of Safety

Verify that the thermal evaluations appropriately address the margins of safety for package temperatures, pressures, and thermal stresses. Verify that these discussions include the effects of uncertainties in thermal properties, test conditions and diagnostics, and analytical methods. If the evaluations are performed by test, verify that the test results are reliable and repeatable.

3.5.4 Thermal Evaluation under Normal Conditions of Transport

3.5.4.1 Heat and Cold

Confirm that the thermal evaluation demonstrates that the tests for normal conditions of transport do not result in significant reduction in packaging effectiveness, including:

- Degradation of the heat-transfer capability of the packaging (such as creation of new gaps between components)
- Changes in material conditions or properties (e.g., expansion, contraction, gas generation, and thermal stresses) that affect the structural performance
- Changes in the packaging that affect containment, shielding, or criticality such as thermal decomposition or melting of materials
- Ability of the packaging to withstand the tests under hypothetical accident conditions.

Verify that the component temperatures and pressures do not exceed their allowable values.

Ensure that the maximum temperature of the accessible package surface is less than 50°C (122°F) for non-exclusive-use shipment or 85°C (185°F) for exclusive use shipment when the package is subjected to the heat conditions of §71.43(g).

3.5.4.2 Maximum Normal Operating Pressure

Confirm that the thermal evaluation determines the maximum normal operating pressure when the package has been subjected to the heat condition for one year. Ensure that the maximum normal operating pressure calculation has considered all possible sources of gases, such as:

- Gases initially present in package
- Saturated vapor, including water vapor from the contents or packaging
- Helium from the radioactive decay of the contents
- Hydrogen or other gases resulting from thermal- or radiation-induced decomposition of materials such as water or plastics.

Ensure that the application demonstrates that hydrogen and other flammable gases comprise less than 5% by volume of the total gas inventory within any confined volume. Confirm that the maximum normal operating pressure is consistent with that in the General Information, Structural Evaluation, and Acceptance Tests and Maintenance Program sections.

3.5.4.3 Maximum Thermal Stresses

Ensure that there is an appropriate evaluation of thermal stresses caused by constrained interfaces among package components resulting from temperature gradients and differential thermal expansions. The evaluation should include the maximum stresses as well as cyclic stresses during the service life of the package.

3.5.5 Thermal Evaluation under Hypothetical Accident Conditions

3.5.5.1 Initial Conditions

Prior to the fire test, the package design must be evaluated for the effects of the drop, crush (if applicable), and puncture tests. Ensure that the initial physical condition of the package design used in the thermal evaluations considers these effects.

Verify that the application justifies the most unfavorable initial conditions of:

- An ambient temperature between -29°C (-20°F) and 38°C (100°F) with no insolation (typically, the temperature will be the latter)
- An internal pressure of the package equal to the maximum normal operating pressure unless a lower internal pressure, consistent with the ambient temperature, is less favorable
- Contents at its maximum decay heat unless a lower heat, consistent with the temperature and pressure, is less favorable.

Confirm that the initial steady-state temperature distribution is consistent with the thermal evaluation under normal conditions of transport.

3.5.5.2 Fire Test Conditions

Confirm that the package design is evaluated for the effects of the fire test. Ensure that the evaluation appropriately addresses:

- Fire dimensions
- Package orientation and support methods
- Fire temperatures and duration
- Heat source
- Availability of adequate oxygen supply.

Verify that after the fire:

- No artificial cooling is applied to the package
- The package is subjected to full insolation
- The evaluation continues until the post-fire, steady-state condition is achieved
- An adequate supply of oxygen is continued throughout this period

- All combustion is allowed to proceed until it terminates naturally.

3.5.5.3 Maximum Temperatures and Pressure

Verify that the evaluation appropriately determines both the transient peak temperatures of package components as a function of time after the fire and the maximum temperatures from the post-fire, steady-state condition. Confirm that these temperatures do not exceed their maximum allowable values. Confirm that lead shielding does not reach melting temperature.

Confirm that the evaluation of the maximum pressure in the package design is based on the maximum normal operating pressure (Section 3.5.4.2) as it is affected by fire-caused increases in package component temperatures. Also verify that possible increases in gas inventory, caused by fire-induced thermal combustion or decomposition processes, have been accounted for in the pressure determination.

Verify that the value of this maximum pressure is consistent with that in the Structural Evaluation and Containment sections.

3.5.5.4 Maximum Thermal Stresses

Ensure that there is an adequate evaluation of thermal stresses caused by constrained interfaces among package components resulting from temperature gradients and differential thermal expansions. Verify that the maximum thermal stresses, which can occur either during or after the fire, are consistent with the Structural Evaluation section.

3.5.6 Appendix

3.5.6.1 Description of Test Facilities

Confirm that the descriptions of a test facility include:

- Type of facility (furnace, pool-fire, etc.)
- Method of heating the package (gas burners, electrical heaters, etc.)
- Volume and emissivity of the furnace interior
- Method of simulating decay heat, if applicable
- Types, locations, and measurement uncertainties of all sensors used to measure the fire heat fluxes, fire temperatures, and test package component temperatures and pressures
- The post-fire environment for a period adequate to attain the post-fire, steady-state condition
- Methods for both maintaining and measuring an adequate supply and circulation of oxygen for both initiating and naturally terminating the combustion of any burnable package component throughout both the fire and post-fire periods.

3.5.6.2 Test Descriptions

Verify that complete descriptions of tests are included in the appendix. This description should include:

- Test procedures

- Test package description
- Test initial and boundary conditions
- Test chronologies (planned and actual)
- Photographs of the package components, including any structural or thermal damage, before and after the tests
- Test measurements, including, at a minimum, documentation of test package physical changes and temperature and heat flux histories
- Corrected test results
- Methods used to obtain these corrected results.

Confirm that all sensors which measure heat fluxes and temperatures are positioned to measure values affecting critical components such as seals, valves, pressure, and structural components. The sensors should have proper operating ranges for the test conditions. Verify that possible perturbations caused by the presence of these sensors (e.g., by disturbing local convective heat transfer conditions) are appropriately considered.

For a pool-fire facility, verify that the fire dimensions and test package relative location conform to the specification in §71.73(c)(4).

- The fire width should extend horizontally between one and four meters beyond any external surface of the package.
- The package should be positioned one meter above the surface of the fuel source.

Since it is probable that the method of supporting the package in the test facility will locally perturb fire conditions adjoining the test package, verify that such an effect has been appropriately incorporated into the thermal evaluation.

3.5.6.3 Applicable Supporting Documents or Specifications

Review any reference documents included in this appendix. In addition to the documents noted in Sections 3.5.6.1 and 3.5.6.2, these may include a variety of items such as thermal specifications of O-rings and other components, and documentation of the thermal properties.

3.5.6.4 Analyses Details

Supplemental calculations may be required to support evaluations presented in the Thermal Evaluation section. Verify that all such special analyses are prepared in a manner consistent with Subsection 3.5.3.1.

3.6 Evaluation Findings

The Safety Evaluation Report should include a finding similar to the following:

Based on review of the statements and representations in the application, the staff concludes that the thermal design has been adequately described and evaluated, and that the thermal performance of the package meets the thermal requirements of 10 CFR Part 71.

4 CONTAINMENT REVIEW

4.1 Review Objective

The objective of this review is to verify that the package design satisfies the containment requirements of 10 CFR Part 71 under normal conditions of transport and hypothetical accident conditions.

4.2 Areas of Review

The description and evaluation of the containment design should be reviewed. The containment review should include the following:

4.2.1 Description of Containment System

- Containment Boundary
- Special Requirements for Plutonium

4.2.2 General Considerations

- Type A Fissile Packages
- Type B Packages
- Combustible-Gas Generation

4.2.3 Containment under Normal Conditions of Transport (Type B Packages)

- Containment Design Criterion
- Demonstration of Compliance with Containment Design Criterion

4.2.4 Containment under Hypothetical Accident Conditions (Type B Packages)

- Containment Design Criterion
- Demonstration of Compliance with Containment Design Criterion

4.2.5 Leakage Rate Tests for Type B Packages

4.2.6 Appendix

4.3 Regulatory Requirements

Regulatory requirements of 10 CFR Part 71 applicable to the containment review are as follows:

- The package design must be described and evaluated to demonstrate that it meets the containment requirements of 10 CFR Part 71. [§71.31(a)(1), §71.31(a)(2), §71.33, §71.35(a)]

- The application must identify established codes and standards applicable to the containment design. [§71.31(c)]

- The package must include a containment system securely closed by a positive fastening device that cannot be opened unintentionally or by a pressure that may arise within the package. [§71.43(c)]

- The package must be made of materials and construction that assure there will be no significant chemical, galvanic, or other reactions. [§71.43(d)]

- Any valve or similar device on the package must be protected against unauthorized operation and, except for a pressure relief valve, must be provided with an enclosure to retain any leakage. [§71.43(e)]

- The package must be designed, constructed, and prepared for shipment to ensure no loss or dispersal of radioactive contents under the tests specified in §71.71 (normal conditions of transport). [§71.43(f), §71.51(a)(1)]

- The package may not incorporate a feature intended to allow continuous venting during transport. [§71.43(h)]

- A Type B package must meet the containment requirements of §71.51(a)(1) for normal conditions of transport and §71.51(a)(2) for hypothetical accident conditions, with no dependence on filters or a mechanical cooling system. [§71.51]

- A package containing plutonium in excess of 0.74 TBq (20 Ci) must satisfy the special containment requirements for plutonium. [§71.63]

4.4 Acceptance Criteria

- The package must satisfy the regulatory requirements listed in Section 4.3.

- The package design must meet the containment requirements of 10 CFR Part 71 under normal conditions of transport and hypothetical accident conditions.

4.5 Review Procedures

The containment review should ensure that the package design has been described and evaluated to meet the containment requirements of 10 CFR Part 71 under normal conditions of transport and hypothetical accident conditions.

The containment review is based in part on the descriptions and evaluations presented in the General Information, Structural Evaluation, and Thermal Evaluation sections of the application. Similarly, results of the containment review are considered in the review of Operating Procedures and Acceptance Tests and Maintenance Program. An example of the information flow for the containment review is shown in Figure 4-1.

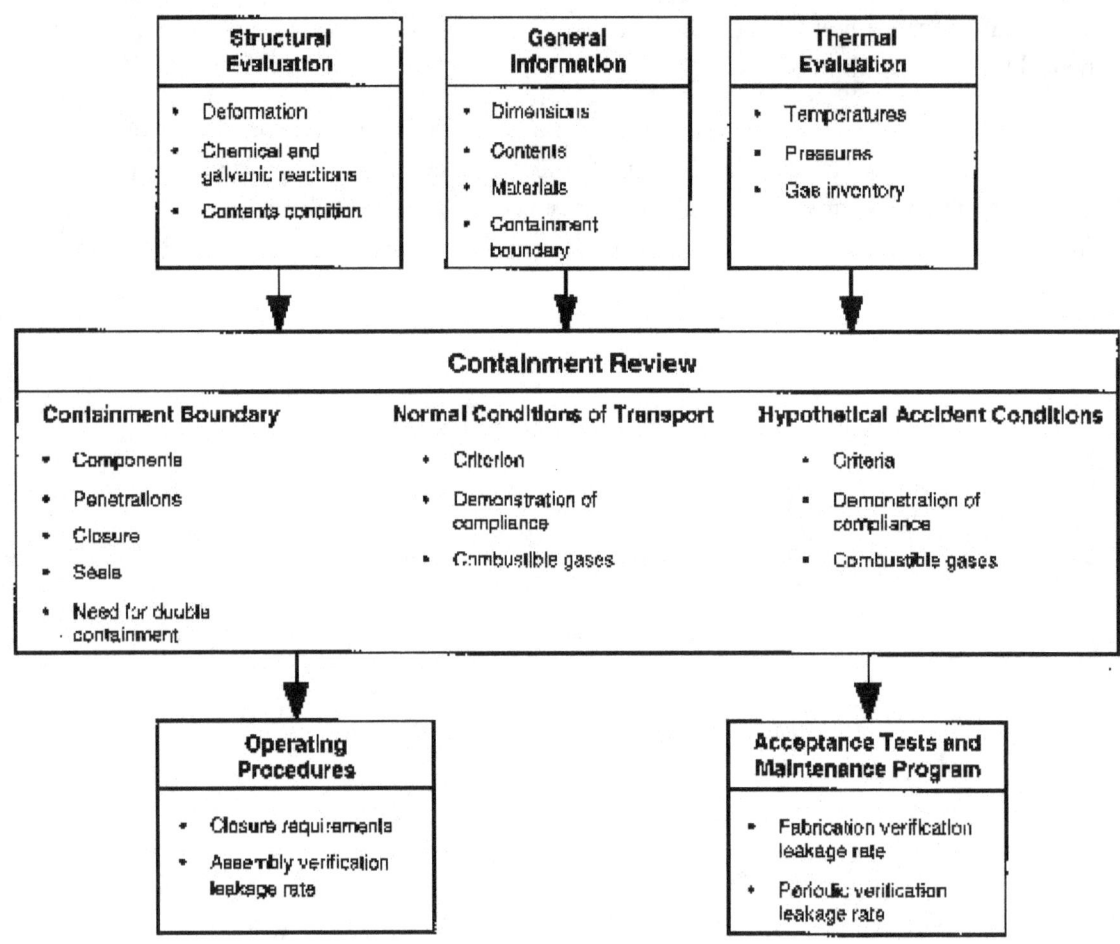

Figure 4-1 Information Flow for the Containment Review

4.5.1 Description of the Containment System

4.5.1.1 Containment Boundary

Review the containment design features presented in the General Information and Containment sections of the application. Verify that the application defines the exact boundary of the containment system. This may include the containment vessel, welds, seals, lids, cover plates, valves, and other closure devices. Ensure that all components of the containment system are shown in the drawings.

Confirm that the following information regarding components of the containment boundary is consistent with that presented in the Structural and Thermal Evaluation sections of the application:

- Materials of construction
- Welds

- Applicable codes and standards (e.g., ASME code specifications for the vessel)
- Bolt torque required to maintain positive closure
- Maximum and minimum allowable temperatures of components, including seals
- Maximum and minimum temperatures of components under the tests for normal conditions of transport and hypothetical accident conditions.

Verify that all containment boundary penetrations and their method of closure are adequately described. Performance specifications for components such as valves and pressure relief devices should be identified, and no device may allow continuous venting. Any valve or similar device on the package must be protected against unauthorized operation and, except for a pressure relief valve, must be provided with an enclosure to retain any leakage. Cover plates and lids should be recessed or otherwise protected. Compliance with the containment requirements specified in 10 CFR Part 71 may not rely on any filter or mechanical cooling system.

Confirm that all containment seals and penetrations, including drain and vent ports, can be leak tested. If fill, drain, or test ports utilize quick-disconnect valves, ensure that such valves do not preclude leakage testing of their containment seals.

Verify that the seal material is appropriate for the package. Ensure that no galvanic, chemical, or other reactions will occur between the seal and the packaging or its contents, and that the seal will not degrade due to irradiation. If penetrations are closed with two seals (e.g., to enable leakage testing), verify which seal is defined as the containment boundary. Ensure that dimensions of the seal grooves are proper for the type and size of seals specified. Confirm that the temperature of containment boundary seals will remain within their specified allowable limits under both normal conditions of transport and hypothetical accident conditions.

Verify that the containment system is securely closed by a positive fastening device that cannot be opened unintentionally or by a pressure that may arise within the package.

4.5.1.2 Special Requirements for Plutonium

If the contents include more than 0.74 TBq (20 Ci) of plutonium, verify that double containment is provided, as specified in §71.63(b), and that plutonium contents are in solid form. Each containment system must meet the requirements of §71.51(a)(1) under normal conditions of transport and §71.51(a)(2) under hypothetical accident conditions.

4.5.2 General Considerations

4.5.2.1 Type A Fissile Packages

For Type A fissile packages, no loss or dispersal of radioactive material is permitted under normal conditions of transport, as specified in §71.43(f). Although 10 CFR Part 71 does not provide numerical release limits (as it does for Type B packages), the package must contain the contents to ensure subcriticality under both normal conditions of transport and hypothetical accident conditions.

4.5.2.2 Type B Packages

Type B packages must satisfy the quantified *release* rates of §71.51. ANSI N14.5 provides an acceptable method to determine the maximum permissible volumetric *leakage* rates based on the allowed regulatory release rates under both normal conditions of transport and hypothetical accident conditions. These two volumetric leakage rates should be converted to standard air leakage rates in accordance with ANSI N14.5. The smaller of these air leakage rates is defined as the reference air leakage rate. Typically, the normal conditions leakage rate is the most restrictive.

Sample analyses for determining containment criteria for Type B packages are provided in NUREG/CR-6487. If the application uses these sample analyses, ensure that the assumptions of that document are applicable to the package under consideration.

4.5.2.3 Combustible-Gas Generation

Confirm that the application demonstrates that any combustible gases generated in the package during a period of one year do not exceed 5% (by volume) of the free gas volume in any confined region of the package. No credit should be taken for getters, catalysts, or other recombination devices.

4.5.3 Containment under Normal Conditions of Transport (Type B Packages)

4.5.3.1 Containment Design Criterion

Confirm that the radionuclides and physical form of the contents evaluated in the Containment section are consistent with those presented in the General Information section of the application. Ensure that the radionuclides include any significant daughter products.

Verify that the application identifies the constituents which comprise the releasable source term, including radioactive gases, liquids, and powder aerosols. If less than 100% of the contents are considered releasable, evaluate the justification for the lower fraction.

Based on the releasable source term, ensure that the maximum permissible release rate and the maximum permissible leakage rate are calculated in accordance with ANSI N14.5. Verify that the maximum normal operating pressure and maximum temperature under normal conditions of transport are consistent with those determined in the Thermal Evaluation section of the application. Using this pressure and temperature, ensure that the maximum permissible leakage rate is converted to the reference air leakage rate in reference cubic centimeters per second (ref cc/s), as defined in ANSI N14.5.

4.5.3.2 Demonstration of Compliance with Containment Design Criterion

Confirm that the application demonstrates that the package meets the containment requirements of §71.51(a)(1) under normal conditions of transport.

- If compliance is demonstrated by test, verify that the leakage rate of a package subjected to the tests of §71.71 does not exceed the maximum allowable leakage rate for normal conditions.

Scale-model testing is not a reliable or acceptable method for quantifying the leakage rate of a full-scale package.

- If compliance is demonstrated by analysis, verify that the structural evaluation shows that the containment boundary, seal region, and closure bolts do not undergo any inelastic deformation and that the materials of the containment system (e.g., seals) do not exceed their maximum allowable temperature limits when subjected to the conditions in §71.71.

- Demonstration that the packaging meets the maximum allowable leakage rate is confirmed during acceptance testing of the packaging, as discussed in the Acceptance Tests and Maintenance Program section.

4.5.4 Containment under Hypothetical Accident Conditions (Type B Packages)

The review procedures for containment under hypothetical accident conditions are similar to those under normal conditions of transport. Differences relevant to hypothetical accident conditions are noted below.

4.5.4.1 Containment Design Criterion

The releasable source term, maximum permissible release rate, and maximum permissible leakage rate should be based on package conditions and the 10 CFR Part 71 containment requirements under hypothetical accident conditions. Verify that the temperatures, pressure, and physical conditions of the package (including the contents) are consistent with those determined in the Structural and Thermal Evaluation sections of the application. Confirm that the maximum permissible leakage rate, when converted to a standard air leakage rate, is greater than the reference air leakage rate.

4.5.4.2 Demonstration of Compliance with Containment Design Criterion

Ensure that the application demonstrates that the package satisfies the containment requirements of §71.51(a)(2) under hypothetical accident conditions. Demonstration is similar to that discussed in Section 4.5.3.2, except that the package should be subjected to the tests of §71.73 and the containment criterion is the maximum allowable leakage rate for hypothetical accident conditions.

4.5.5 Leakage Rate Tests for Type B Packages

Using the reference air leakage rate, confirm that the allowable leakage rates for the following conditions are determined in accordance with ANSI N14.5:

- Fabrication leakage rate test
- Maintenance leakage rate test
- Periodic leakage rate test
- Preshipment leakage rate test.

Fabrication, maintenance, and periodic leakage rate tests should be included in the Acceptance Tests and Maintenance Program review. The preshipment leakage rate test for assembly verification should be included in the Operating Procedures review.

4.5.6 Appendix

Confirm that the appendix includes a list of references, copies of applicable references if not generally available to the reviewer, test results, and other appropriate supplemental information.

4.6 Evaluation Findings

The Safety Evaluation Report should include a finding similar to the following:

> Based on review of the statements and representations in the application, the staff concludes that the containment design has been adequately described and evaluated and that the package design meets the containment requirements of 10 CFR Part 71.

4.7 References

American National Standards Institute, ANSI N14.5-1997, "American National Standard for Radioactive Materials–Leakage Tests on Packages for Shipment," New York.

U.S. Nuclear Regulatory Commission, "Containment Analysis for Type B Packages Used to Transport Various Contents," NUREG/CR-6487, November 1996.

5 SHIELDING REVIEW

5.1 Review Objective

The objective of this review is to verify that the package design meets the external radiation requirements of 10 CFR Part 71 under normal conditions of transport and hypothetical accident conditions.

5.2 Areas of Review

The description and evaluation of the shielding design should be reviewed. The shielding review should include the following:

5.2.1 Description of Shielding Design

- Design Features
- Summary Table of Maximum Radiation Levels

5.2.2 Radiation Source

- Gamma Source
- Neutron Source

5.2.3 Shielding Model

- Configuration of Source and Shielding
- Material Properties

5.2.4 Shielding Evaluation

- Methods
- Input and Output Data
- Flux-to-Dose-Rate Conversion
- External Radiation Levels

5.2.5 Appendix

5.3 Regulatory Requirements

Regulatory requirements of 10 CFR Part 71 applicable to the shielding review are as follows:

- The package design must be described and evaluated to demonstrate that it meets the shielding requirements of 10 CFR Part 71. [§71.31(a)(1), §71.31(a)(2), §71.33, §71.35(a)]

- The application must identify the established codes and standards used for the shielding design. [§71.31(c)]

- The package must be designed, constructed, and prepared for shipment so that the external radiation levels will not significantly increase under the tests specified in §71.71 (normal conditions of transport). [§71.43(f), §71.51(a)(1)]

- Under the tests specified in §71.71 (normal conditions of transport), the external radiation levels must meet the requirements of §71.47(a) for non-exclusive-use or §71.47(b) for exclusive-use shipments. [§71.47]

- Under the tests specified in §71.73 (hypothetical accident conditions), the external radiation level must not exceed 10 mSv/h (1 rem/h) at one meter from the surface of the package. [§71.51(a)(2)]

5.4 Acceptance Criteria

- The package must satisfy the regulatory requirements listed in Section 5.3.

- The package design must meet the external radiation requirements of 10 CFR Part 71 under normal conditions of transport and hypothetical accident conditions.

5.5 Review Procedures

The shielding review should ensure that the package design has been described and evaluated to meet the external radiation requirements of 10 CFR Part 71 under normal conditions of transport and hypothetical accident conditions.

The shielding review is based in part on the descriptions and evaluations presented in the General Information, Structural Evaluation, and Thermal Evaluation sections of the application. Results of the shielding review are considered in the review of Operating Procedures and the Acceptance Tests and Maintenance Program. An example of this information flow for the shielding review is shown in Figure 5-1.

5.5.1 Description of Shielding Design

5.5.1.1 Design Features

Review the shielding design features presented in the General Information and Shielding Evaluations sections of the application. Design features important to shielding include:

- Dimensions, tolerances, and densities of material for neutron or gamma shielding, including those packaging components considered in the shielding evaluation
- Mass density, atomic density, or areal density of materials used as neutron absorbers
- Structural components that maintain the contents in a fixed position within the package
- Dimensions of the transport vehicle that are considered in the shielding evaluation.

Confirm that the text and sketches describing the shielding design features are consistent with the engineering drawings and the models used in the shielding evaluation.

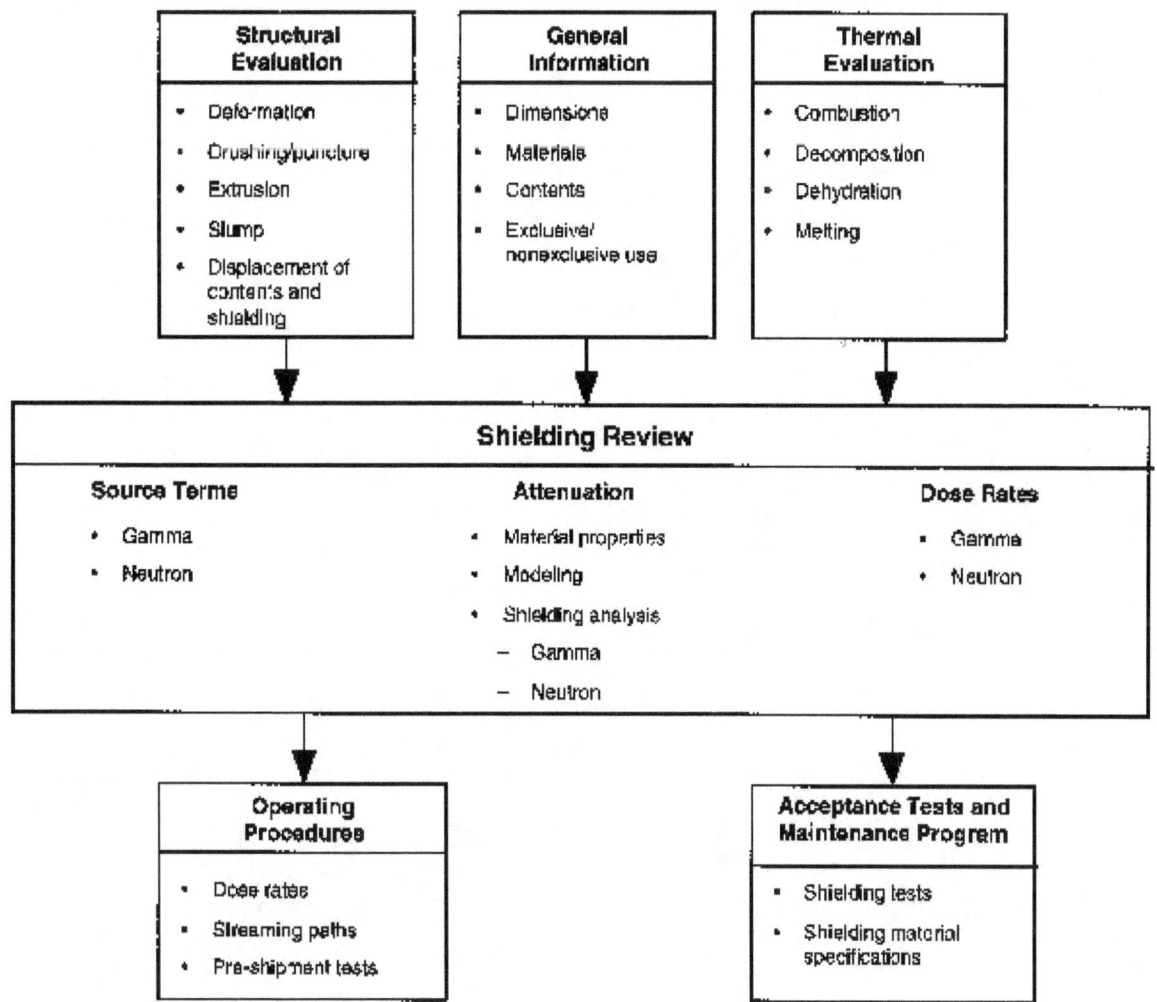

Figure 5-1 Information Flow for the Shielding Review

5.5.1.2 Summary Table of Maximum Radiation Levels

Review the summary table of maximum radiation levels. Ensure that the maximum dose rates are presented for both normal conditions of transport and hypothetical accident conditions at the appropriate locations for non-exclusive or exclusive use (or both), as applicable. Table 5.1 is an example of the information that should be presented for non-exclusive use.

Verify that the radiation levels are within the regulatory limits as indicated in Table 5.2. Examine the variation of dose rates at different package locations for general consistency. Confirm that dose rates decrease as either the distance from the source or as the shielding effectiveness (e.g., thickness) increases.

Table 5.1 Example for Summary Table of External Radiation Levels (Non-Exclusive Use)

Normal Conditions of Transport	Package Surface mSv/h (mrem/h)			1 Meter from Package Surface mSv/h (mrem/h)		
Radiation	Top	Side	Bottom	Top	Side	Bottom
Gamma						
Neutron						
Total						
10 CFR 71.47(a) Limit	2 (200)	2 (200)	2 (200)	0.1 (10)*	0.1 (10)*	0.1 (10)*

* Transport index may not exceed 10

Hypothetical Accident Conditions	1 Meter from Package Surface mSv/h (mrem/h)		
Radiation	Top	Side	Bottom
Gamma			
Neutron			
Total			
10 CFR 71.51(a) Limit	10 (1000)	10 (1000)	10 (1000)

Table 5.2 Package and Vehicle Radiation Level Limits[a]

Note: This table is not a substitute for NRC or DOT regulations on transportation of radioactive materials.

Transport Vehicle Use:	Non-Exclusive	Exclusive		
Transport Vehicle Type:	Open or Closed	Open (flat-bed)	Open with Enclosure[b]	Closed
Package (or freight container) Limits:				
External Surface	2 mSv/h (200 mrem/h)	2 mSv/h (200 mrem/h)	10 mSv/h (1000 mrem/h)	10 mSv/h (1000 mrem/h)
1 meter from External Surface[c]	0.1 mSv/h (10 mrem/h)	No Limit		
Roadway or Railway Vehicle (or freight container) Limits:				
Any point on outer surface	N/A	N/A	N/A	2 mSv/h (200 mrem/h)
Vertical planes projected from outer edges	N/A	2 mSv/h (200 mrem/h)	2 mSv/h (200 mrem/h)	N/A
Top of...	N/A	Load: 2 mSv/h (200 mrem/h)	Enclosure: 2 mSv/h (200 mrem/h)	Vehicle: 2 mSv/h (200 mrem/h)
2 meters from...	N/A	Vertical Planes: 0.1 mSv/h (10 mrem/h)	Vertical Planes: 0.1 mSv/h (10 mrem/h)	Outer Lateral Surfaces: 0.1 mSv/h (10 mrem/h)
Underside	N/A	2 mSv/h (200 mrem/h)		
Occupied position	N/A	0.02 mSv/h (2 mrem/h)[d]		

[a] The limits in this table do not apply to excepted packages see 49 CFR 173.421-426.
[b] Securely attached (to vehicle), access-limiting enclosure; package personnel barriers are considered as enclosures.
[c] Transport index may not exceed 10.
[d] Does not apply to private carrier wearing dosimetry if under radiation protection program satisfying 10 CFR Part 20.

5.5.2 Radiation Source

Confirm that the contents used in the shielding analysis are consistent with those specified in the General Information section of the application. If the package is designed for multiple types of contents, ensure that the contents producing the highest external dose rate at each location are clearly identified and evaluated.

5.5.2.1 Gamma Source

Based on the contents specified, verify that the maximum gamma source strength and spectra are calculated by an appropriate method (e.g., standard computer codes and hand calculations). Ensure that the source contribution from radioactive daughter products is included if it produces higher dose rates than the contents without decay. If the radioactive nuclides and gamma spectra are calculated with a computer code, review the key parameters described in the application or listed in the input file. Verify that the production of secondary gammas (e.g., from (n,?) reactions in shielding material) is either calculated as part of the shielding evaluation (see Section 5.5.4.1) or otherwise appropriately included in the source term.

Confirm that the results of the source term determination are presented as a listing of gammas per second, or MeV per second, as a function of energy. The activity (or mass) of each nuclide that contributes significantly to the source term should also be provided as supporting information. The review should independently confirm the source term specified in the application.

5.5.2.2 Neutron Source

Review the method used to determine the neutron source term. Verify that the method considers, as appropriate, neutrons from both spontaneous fission and from (a,n) reactions. If the application assumes that either of these source contributions is negligible, ensure that an appropriate justification is provided. Verify that the production of neutrons from subcritical multiplication is either calculated as part of the shielding evaluation (see Section 5.5.4.1) or otherwise appropriately included in the source term.

Confirm that the results of the source term calculation, if applicable, are presented as a listing of neutrons per second as a function of energy. The contribution from spontaneous fission and (a,n) should be separately identified, along with the actinides or light nuclei significant for these processes. For packages with significant neutron sources, the review should independently confirm the neutron source term.

5.5.3 Shielding Model

Review the Structural and Thermal Evaluation sections of the application to determine the effects that the tests for normal conditions of transport and hypothetical accident conditions have on the packaging and its contents. Verify that the models used in the shielding calculation are consistent with these effects.

5.5.3.1 Configuration of Source and Shielding

Verify the dimensions of the source and packaging used in the shielding models. If contents can be positioned at varying locations or with varying densities, ensure that the location and physical properties of the contents used in the evaluation are those resulting in the maximum external radiation levels. For example, the source configuration that maximizes radiation level on the side of the package might not be the same source configuration that maximizes the radiation level on the top or bottom. Ensure that any changes in configuration (e.g., displacement of source or shielding, reduction in shielding) resulting under normal conditions of transport or hypothetical accident conditions have been included, as appropriate.

For exclusive-use shipments in which the analysis is based on the radiation levels of §71.47(b), confirm that dimensions of the transport vehicle and package location are included as appropriate.

Verify that the dose point locations in the shielding model include all locations prescribed in §§71.47(a) or 71.47(b), and §71.51(a)(2). Ensure that these points are chosen to identify the location of the maximum radiation levels. Confirm that voids, streaming paths, and irregular geometries are included in the model or otherwise treated in an adequate manner.

5.5.3.2 Material Properties

Verify the appropriate material properties (e.g., mass densities and atom densities) used in the shielding models of the packaging, contents, and conveyance (if applicable). Ensure that any changes resulting under normal conditions of transport or hypothetical accident conditions have been included, as appropriate. Melting of lead shielding is not acceptable. Confirm that shielding properties will not degrade during the service life of the packaging (e.g., degradation of foam or dehydration of hydrogenous materials).

If the shielding model considers a homogenous source region (rather than a detailed heterogeneous model of the contents), ensure that such an approach is justified, and verify that the homogenized mass densities are correct for normal conditions of transport and hypothetical accident conditions. Atom densities should also be confirmed if used as input to shielding calculations.

5.5.4 Shielding Evaluation

5.5.4.1 Methods

Ensure that the methods used for the shielding evaluation are appropriate. Standard computer programs should be referenced. Other codes or methods should be described in the application, and appropriate supplemental information should be provided. Verify that the number of dimensions of the code is appropriate for the package geometry, including streaming paths, if applicable.

Confirm that the cross-section library used by the code is applicable for shielding calculations. Ensure that the code accounts for subcritical multiplication and secondary gamma production unless these conditions have been otherwise appropriately considered (e.g., in the source-term specification).

5.5.4.2 Input and Output Data

Verify that key input data for the shielding calculations are identified. These will depend on the type of code (e.g., deterministic or Monte Carlo), as well as the code itself. The application should also include representative input files used in the analyses. Verify, as appropriate, that the information from the shielding models is properly input into the code.

At least one representative output file (or key sections of the file) should generally be included in the application. Ensure that proper convergence is achieved and that the calculated dose rates from the output files agree with those reported in the text.

5.5.4.3 Flux-to-Dose-Rate Conversion

Ensure that the evaluation properly converts the gamma and neutron flux to dose rates. This conversion should generally use ANSI/ANS 6.1.1-1977 although other conversions may be used for point-kernel gamma calculations. Use of the conversions in ANSI/ANS 6.1.1-1991 can result in a significant underestimation of external dose rates (as defined by 49 CFR 173.403 and 10 CFR 20.1004). In addition, the dose rates determined with the 1991 standard do not correspond physically to dose rates measured by typical radiation monitoring instruments.

Verify the accuracy of the flux-to-dose rate conversion factors, which should be tabulated as a function of the energy group structure used in the shielding calculation.

5.5.4.4 External Radiation Levels

Confirm that the external radiation levels under normal conditions of transport and hypothetical accident conditions agree with the summary tables discussed in Section 5.5.1.2 and that they meet the limits in §71.47(a) or §71.47(b), as appropriate, and §71.51(a)(2). Verify that the analysis shows that the locations selected are those of maximum dose rates. To determine maximum dose rates, radiation levels may be averaged over the cross-sectional area of a probe of reasonable size (NUREG/CR-5569). For packages with streaming paths or voids, averaging should not be used to reduce the radiation levels resulting from such features.

Ensure that the external radiation levels are reasonable and that their variations with location are consistent with the geometry and shielding characteristics of the package.

Confirm that the evaluation addresses damage to the shielding under normal conditions of transport and hypothetical accident conditions. Verify that any damage under normal conditions of transport (§71.71) does not result in a significant increase in the external dose rates, as required by §71.43(f) and §71.51(a)(1). Any increase should be explained and justified as not significant.

The review should include a confirmatory analysis of the shielding calculations reported in the application. Because measurements of the actual dose rates from packages must be performed prior to shipment in order to show that the §71.47 limits are satisfied, a number of factors should be considered in determining the level of effort of the confirmatory analysis. These factors include the expected magnitude of the dose rates, margin between calculations and regulatory limits, similarity with previously reviewed packages, thoroughness of the review of source terms and other input data, and bounding assumptions in the analysis.

5.5.5 Appendix

Confirm that the appendix includes a list of references, copies of applicable references if not generally available, computer code descriptions, input and output files, test results, and other appropriate supplemental information.

5.6 Evaluation Findings

The Safety Evaluation Report should include a finding similar to the following:

Based on review of the statements and representations in the application, the staff concludes that the shielding design has been adequately described and evaluated and that the package meets the external radiation requirements of 10 CFR Part 71.

5.7 References

American Nuclear Society, "American National Standard for Neutron and Gamma-Ray Fluence to Dose Factors," ANSI/ANS 6.1.1-1991, LaGrange Park, Illinois.

American Nuclear Society, "American National Standard for Neutron and Gamma-Ray Flux to Dose Rate Factors," ANSI/ANS 6.1.1-1977, LaGrange Park, Illinois.

Broadhead, B. L., C. V. Parks, and R. B. Pope, "Assessment of Proposed Dose Factor Changes to Shipping Cask Design and Operation," *Proc. of the Third International Conference on High Level Radioactive Waste Management*, Las Vegas, Nevada, 1992.

U.S. Nuclear Regulatory Commission, "Averaging of Radiation Levels Over the Detector Probe Area, HPPOS-13, in *Health Physics Positions Data Base*, NUREG/CR-5569, Rev. 1, 1992.

6 CRITICALITY REVIEW

6.1 Review Objective

The objective of this review is to verify that the package design meets the criticality safety requirements of 10 CFR Part 71 under normal conditions of transport and hypothetical accident conditions.

6.2 Areas of Review

The description and evaluation of the criticality design should be reviewed. The criticality review should include the following:

6.2.1 Description of Criticality Design

- Design Features
- Summary Table of Criticality Evaluations
- Transport Index

6.2.2 Fissile Material Contents

6.2.3 General Considerations

- Model Configuration
- Material Properties
- Computer Codes and Cross-Section Libraries
- Demonstration of Maximum Reactivity
- Confirmatory Analyses

6.2.4 Single Package Evaluation

- Configuration
- Results

6.2.5 Evaluation of Package Arrays under Normal Conditions of Transport

- Configuration
- Results

6.2.6 Evaluation of Package Arrays under Hypothetical Accident Conditions

- Configuration
- Results

6.2.7 Benchmark Evaluations

- Applicability of Benchmark Experiments
- Bias Determination

6.2.8 Appendix

6.3 Regulatory Requirements

Regulatory requirements of 10 CFR Part 71 applicable to the criticality review of fissile material packages are as follows:

- The package design must be described and evaluated to demonstrate that it meets the criticality requirements of 10 CFR Part 71. [§71.31(a)(1), §71.31(a)(2), §71.33, §71.35(a)]

- The application must identify the established codes and standards used for the criticality design. [§71.31(c)]

- Unknown properties of fissile material must be assumed to be those which will result in the highest neutron multiplication. [§71.83]

- The package must be designed, constructed, and prepared for shipment so that there will be no significant reduction in the effectiveness of the packaging under the tests specified in §71.71 (normal conditions of transport). [§71.43(f), §71.51(a)(1), §71.55(d)(4)]

- A single package must meet the requirements of §71.55(d) under normal conditions of transport.

- A single package must meet the requirements of §71.55(e) under hypothetical accident conditions.

- A single package must be designed and constructed and its contents limited so that it would be subcritical if water were to leak into the containment system. [§71.55(b)]

- An array of packages must be subcritical under normal conditions of transport and hypothetical accident conditions. [§71.59(a)]

- A fissile material package must be assigned a transport index for nuclear criticality control to limit the number of packages in a single shipment. [§71.59, §71.35(b)]

6.4 Acceptance Criteria

- The package must satisfy the regulatory requirements listed in Section 6.3.

- The package design must meet the nuclear criticality safety requirements of 10 CFR Part 71 under normal conditions of transport and hypothetical accident conditions.

6.5 Review Procedures

The criticality review should ensure that the package design has been described and evaluated to meet the requirements for nuclear criticality safety of 10 CFR Part 71 under normal conditions of transport and hypothetical accident conditions.

The criticality review is based in part on the descriptions and evaluations presented in the General Information, Structural Evaluation, and Thermal Evaluation sections of the application. Results of the criticality review are considered in the review of the Operating Procedures and the Acceptance Tests and Maintenance Program. An example of this information flow for the criticality review is shown in Figure 6-1.

6.5.1 Description of Criticality Design

6.5.1.1 Design Features

Review the General Information section of the application and any additional description of the criticality design presented in the Criticality Evaluation section of the application. Design features important for criticality include:

- Dimensions and tolerances of the containment system for fissile material
- Structural components that maintain the fissile material or neutron poisons in a fixed position within the package or in a fixed position relative to each other
- Location, dimensions, and concentration of neutron absorbing materials and moderating materials, including neutron poisons and shielding material
- Dimensions and tolerances of floodable voids and flux traps within the package
- Dimensions and tolerances of the overall package that affect the physical separation of the fissile material contents in package arrays.

Confirm that the text and sketches describing the criticality design features are consistent with the engineering drawings and the models used in the criticality evaluation.

6.5.1.2 Summary Table of Criticality Evaluation

Review the summary table of the criticality evaluation, which should address the following cases, as described in Sections 6.5.4 through 6.5.6:

- A single package, under the conditions of §71.55(b), (d), and (e)
- An array of undamaged packages, under the conditions of §71.59(a)(1)
- An array of damaged packages, under the conditions of §71.59(a)(2).

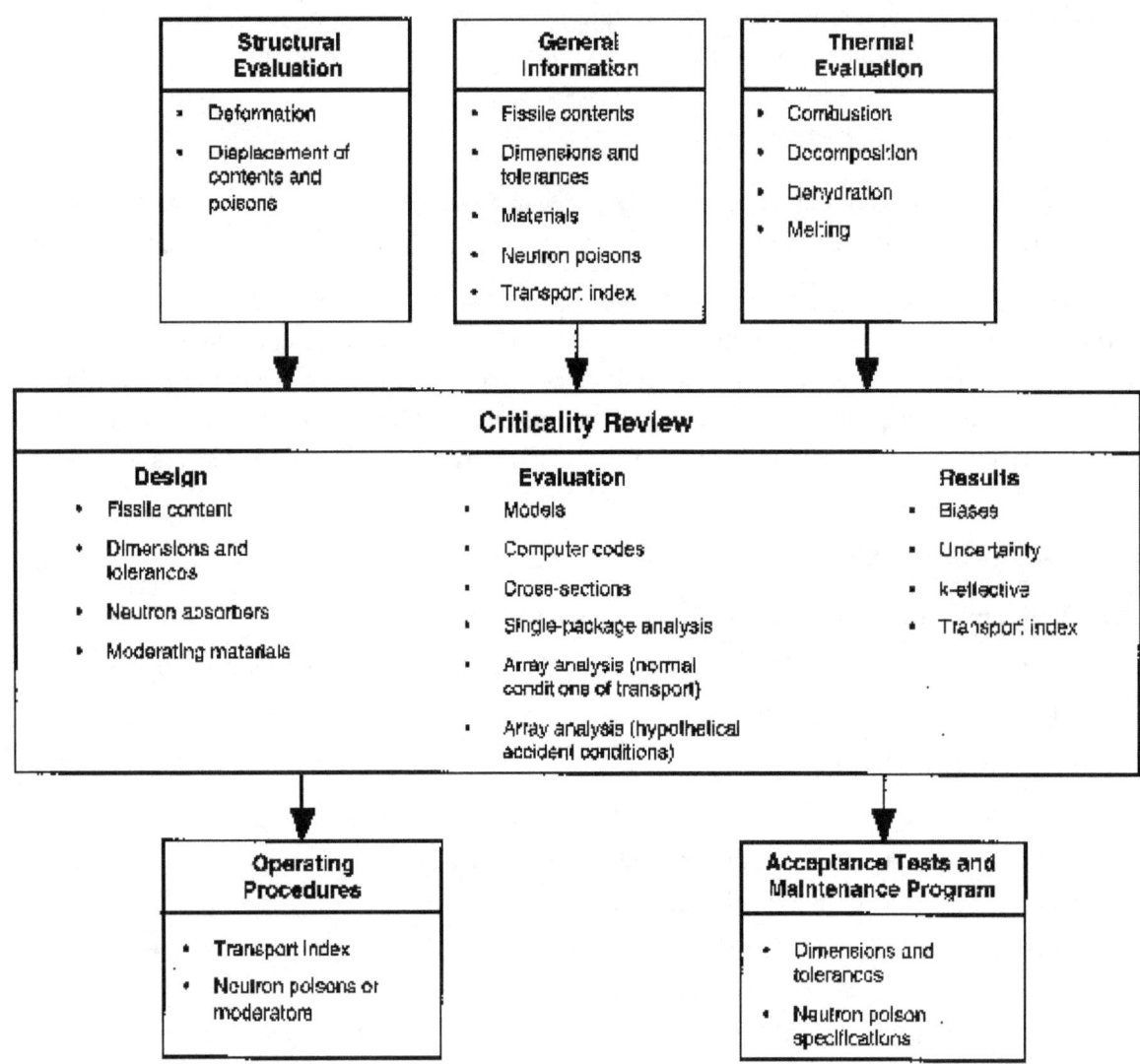

Figure 6-1 Information Flow for the Criticality Review

Verify that the table includes the maximum value of the effective multiplication factor (k_{eff}), the uncertainty, the bias, and the number of packages evaluated in the arrays. The table should also show that the sum of k_{eff}, two standard deviations, and the bias adjustment does not exceed 0.95 for each case.

Confirm that the summary table illustrates that the package meets the above subcriticality criterion.

6.5.1.3 Transport Index

Based on the number of packages evaluated in the arrays, verify that the application determines the appropriate N and calculates the criticality transport index correctly. Ensure that this transport index is consistent with that reported in the General Information section of the application.

6.5.2 Fissile Material Contents

Ensure that the specifications for the contents used in the criticality evaluation are consistent with those in the General Information section of the application. Specifications relevant to the criticality evaluation include fissile material mass, dimensions, enrichment, physical and chemical composition, density, moisture, and other characteristics dependent on the specific contents. Any differences from the specifications in the General Information section should be clearly identified and justified. Because a partially filled container may allow more room for moderators (e.g., water), the most reactive case may be for a mass of fissile material that is less than the maximum allowable contents.

If the package is designed for multiple types of contents, the application may include a separate criticality evaluation and propose different criticality controls for each content type. Any assumptions that certain contents need not be evaluated because they are less reactive than evaluated contents should also be properly justified. The review procedures of this section are applicable for each content type evaluated in the application.

6.5.3 General Considerations

The considerations discussed below are applicable to the criticality evaluations of a single package and arrays of packages under normal conditions of transport and hypothetical accident conditions.

General guidance for preparing criticality evaluations of transportation packages is provided in NUREG/CR-5661.

6.5.3.1 Model Configuration

Examine the Structural and Thermal Evaluation sections of the application to determine the effects of the normal conditions of transport and hypothetical accident conditions on the packaging and its contents. Verify that the models used in the criticality calculation are consistent with these effects.

Verify the dimensions of the contents and packaging used in the criticality models. For some types of packagings and contents, e.g., powders, the contents can be positioned at varying locations and densities. The relative location and physical properties of the contents within the packaging should be justified as those resulting in the maximum multiplication factor. Dimensional tolerances, e.g., for cavity sizes and poison thickness, should be considered in the manner which maximizes reactivity.

6.5.3.2 Material Properties

Verify that the appropriate mass densities and atom densities are provided for materials used in the models of the packaging and contents. Material properties should be consistent with the condition of the package under the tests of §71.71 and §71.73, and any differences between normal conditions of transport and hypothetical accident conditions should be addressed.

Ensure that materials relevant to the criticality design (e.g., poisons, foams, plastics, and other hydrocarbons) are properly specified. No more than 75% of the specified minimum neutron poison concentration should generally be considered in the criticality evaluation. Verify that materials will not degrade during the service life of the packaging.

6.5.3.3 Computer Codes and Cross-Section Libraries

Verify that the application uses an appropriate computer code (or other acceptable method) for the criticality evaluation. Standard codes should be clearly referenced. Other codes or methods should be described in the application, and appropriate supplemental information should be provided.

Ensure that the criticality evaluations use an appropriate cross-section library. If multigroup cross sections are used, confirm that the neutron spectrum of the package has been appropriately considered and that the cross sections are properly processed to account for resonance absorption and self-shielding. Additional information regarding cross-sections is provided in NMSS Information Notice No. 91-26 and NUREG/CR-6328.

Verify that the code has been properly used in the criticality evaluation. Key input data for the criticality calculations should be identified. These include number of neutrons per generation, number of generations, convergence criteria, mesh selection, etc., depending on the code used. The application should include at least one representative input file for a single package, undamaged array, and damaged array evaluation. Verify, as appropriate, that the information from the criticality model, material properties, and cross sections is properly input into the code.

At least one representative output file (or key sections) should be included in the application. Ensure that the calculation has properly converged and that the calculated multiplication factors from the output files agree with those reported in the evaluation.

6.5.3.4 Demonstration of Maximum Reactivity

Verify that the analyses demonstrate the most reactive configuration of each case listed in Section 6.5.1.2 (single package, array of undamaged packages, and array of damaged packages). Assumptions and approximations should be clearly identified and justified.

Ensure that the analysis determines the optimum combination of internal moderation (within the package) and interspersed moderation (between packages), as applicable. Confirm that preferential flooding of different regions within the package is considered as appropriate. As noted in Section 6.5.2, the maximum allowable fissile material is not necessarily the most reactive contents.

Additional guidance on determining the most reactive configurations is presented in NUREG/CR-5661.

6.5.3.5 Confirmatory Analyses

The review should include a confirmatory analysis of the criticality calculations reported in the application. As a minimum, perform an independent calculation of the most reactive case, as well as sensitivity analyses to confirm that the most reactive case has been correctly identified. To the extent practical, use an independent model of the package and a different code and cross-section set from that used in the application.

6.5.4 Single Package Evaluation

6.5.4.1 Configuration

Ensure that the criticality evaluation demonstrates that a single package is subcritical under both normal conditions of transport and hypothetical accident conditions. The evaluations should consider:

- Fissile material in its most reactive credible configuration consistent with the condition of the package and the chemical and physical form of the contents
- Water moderation to the most reactive credible extent, including water inleakage to the containment system as specified in §71.55(b)
- Full water reflection on all sides of the containment system as specified in §71.55(b)(3), or reflection by the package materials, whichever results in the maximum reactivity.

6.5.4.2 Results

Verify that the package also meets the additional specifications of §§71.55(d)(2) through 71.55(d)(4) under normal conditions of transport.

Ensure that the results of the most reactive case for the single package analysis are consistent with the information presented in the summary table discussed in Section 6.5.1.2. If the package can be shown to be subcritical by reference to a standard such as ANSI/ANS 8.1 (in lieu of calculations), verify that the standard is applicable to the package conditions.

6.5.5 Evaluation of Package Arrays under Normal Conditions of Transport

6.5.5.1 Configuration

Ensure that the criticality evaluation demonstrates that an array of 5N packages is subcritical under normal conditions of transport. The evaluation should consider:

- The most reactive configuration of the array (e.g., pitch and package orientation) with nothing between the packages
- The most reactive credible configuration of the packaging and its contents under normal conditions of transport. If the water spray test has demonstrated that water would not leak into the package, water inleakage need not be assumed.

- Full water reflection on all sides of a finite array.

6.5.5.2 Results

Verify that the most reactive array conditions are clearly identified and that the results of the analysis are consistent with the information presented in the summary table discussed in Section 6.5.1.2.

Confirm that the appropriate N value is used to determine the transport index. The appropriate N should be the smaller value which assures subcriticality for 5N packages under normal conditions of transport or 2N packages under hypothetical accident condition, as discussed in the next section.

6.5.6 Evaluation of Package Arrays under Hypothetical Accident Conditions

6.5.6.1 Configuration

Ensure that the criticality evaluation demonstrates that an array of 2N packages is subcritical under hypothetical accident conditions. The evaluation should consider:

- The most reactive configuration of the array (e.g., pitch, package orientation, and internal moderation)
- Optimum interspersed hydrogenous moderation
- The most reactive credible configuration of the packaging and its contents under hypothetical accident conditions, including inleakage of water
- Full water reflection on all sides of a finite array.

6.5.6.2 Results

Verify that the most reactive array conditions are clearly identified and that the results of the analysis are consistent with the information presented in the summary table discussed in Section 6.5.1.2.

Confirm that the appropriate N value is used to determine the transport index. The appropriate N should be the smaller value which assures subcriticality for 2N packages under hypothetical accident conditions or 5N packages under normal conditions of transport, consistent with the previous section.

6.5.7 Benchmark Evaluations

Ensure that the computer codes for criticality calculations are benchmarked against critical experiments. Verify that the analysis of the benchmark experiments uses the same computer code, hardware, and cross-section library as those used to calculate the k_{eff} values for the package.

Additional guidance on benchmarking of nuclear criticality codes is provided in NUREG/CR-6361.

6.5.7.1 Applicability of Benchmark Experiments

Review the general description of the benchmark experiments and confirm that they are appropriately referenced.

Verify that the benchmark experiments are applicable to the actual packaging design and contents. The benchmark experiments should have, to the maximum extent possible, the same materials, neutron spectra, and configuration as the package evaluations. Key package parameters that should be compared with those of the benchmark experiments include type of fissile material, enrichment, H/U ratio, poison, and configuration. Confirm that differences between the package and benchmarks are identified and properly considered.

In addition, the application should address the overall quality of the benchmark experiments and the uncertainties in experimental data (e.g., mass, density, dimensions). Ensure that these uncertainties are treated in a conservative manner, i.e., they result in a lower multiplication factor for the benchmark experiment.

6.5.7.2 Bias Determination

Examine the results of the calculations for the benchmark experiments and the method used to account for biases, including the contribution from uncertainties in experimental data.

Ensure that a sufficient number of appropriate benchmark experiments are analyzed and that the results of these benchmark calculations are used to determine an appropriate bias for the package calculations. Confirm that the benchmark evaluations address trends in the bias with respect to parameters such as pitch-to-rod diameter, assembly separation, neutron absorber material, etc. Verify that only negative biases (results that under-predict k_{eff}) are considered, with positive bias results treated as zero bias. Additional information on determining a bias and its range of applicability is provided in NUREG/CR-5661 and NUREG/CR-6361.

Statistical and convergence uncertainties of benchmark calculations should also be addressed. Considering the current availability of computer resources, the reviewer should ensure that these uncertainties do not significantly affect the results.

6.5.8 Appendix

Confirm that the appendix includes a list of references, copies of applicable references if not generally available to the reviewer, computer code descriptions, input and output files, test results, and any other appropriate supplemental information.

6.6 Evaluation Findings

The Safety Evaluation Report should include a finding similar to the following:

> Based on review of the statements and representations in the application, the staff concludes that the nuclear criticality safety design has been adequately described and evaluated and that the package meets the subcriticality requirements of 10 CFR Part 71.

6.7 References

American Nuclear Society, "American National Standard for Nuclear Criticality Safety in Operations with Fissionable Material Outside Reactors," ANSI/ANS 8.1-1983 (R1988), LaGrange Park, Illinois.

U.S. Nuclear Regulatory Commission, "Adequacy of the 123-Group Cross-Section Library for Criticality Analyses of Water-Moderated Uranium Systems," NUREG/CR-6328, August 1995.

U.S. Nuclear Regulatory Commission, "Criticality Benchmark Guide for Light-Water-Reactor Fuel in Transportation and Storage Packages," NUREG/CR-6361, January 1997.

U.S. Nuclear Regulatory Commission, "Potential Nonconservative Errors in the Working Format Hansen-Roach Cross-Section Set Provided with the KENO and SCALE Codes, " NMSS Information Notice No. 91-26, April 15, 1991.

U.S. Nuclear Regulatory Commission, "Recommendations for Preparing the Criticality Safety Evaluation of Transportation Packages," NUREG/CR-5661, April 1997.

7 OPERATING PROCEDURES REVIEW

7.1 Review Objective

The objective of this review is to verify that the operating controls and procedures meet the requirements of 10 CFR Part 71 and that the operating procedures are adequate to assure the package will be operated in a manner consistent with its evaluation for approval.

7.2 Areas of Review

Procedures which assure that the package will be operated in a manner consistent with its evaluation for approval should be reviewed. The operating procedures review should include the following:

7.2.1 Package Loading

- Preparation for Loading
- Loading of Contents
- Preparation for Transport

7.2.2 Package Unloading

- Receipt of Package from Carrier
- Removal of Contents

7.2.3 Preparation of Empty Package for Transport

7.2.4 Other Procedures

7.2.5 Appendix

7.3 Regulatory Requirements

Regulatory requirements of 10 CFR Part 71 applicable to the operating procedures review are as follows:

- The application must identify the established codes and standards used for the operating procedures. [§71.31(c)]

- The application for a fissile material shipment must include any special controls and precautions for transport, loading, unloading, and handling and any special controls in case of accident or delay. [§71.35(c)]

- Packages that require exclusive-use shipment because of increased radiation levels must be controlled by providing written instructions to the carrier. [§§71.47(b-d)]

- Before each shipment, the licensee must ensure that the package meets the routine-determination requirements of 10 CFR Part 71. [§71.87]

- Prior to delivery of a package to a carrier, the licensee must send any special instructions needed to safely open the package to the consignee for the consignee's use in accordance with 10 CFR 20.1906(e). [§71.89]

7.4 Acceptance Criteria

- The operating procedures must meet the regulatory requirements listed in Section 7.3.

- The operating procedures must be adequate to assure that the package will be operated in a manner consistent with the basis used for its safety evaluation.

7.5 Review Procedures

The review should verify that the operating controls and procedures meet the requirements of 10 CFR Part 71, and that these procedures are adequate to assure the package will be operated in a manner consistent with its evaluation for approval. Detailed procedures that could be implemented without further expansion are not required.

The commitments specified in the Operating Procedures section of the application are typically included by reference into the certificate of compliance as conditions of the package approval. Package operation and preparation for shipment must be performed in accordance with detailed written procedures. The operating procedures submitted as part of the application should establish the minimum steps necessary to assure safe performance of the package under normal conditions of transport and hypothetical accident conditions. These steps should be presented in sequential order, as applicable.

The operating procedures review is based in part on the descriptions and evaluations presented in the General Information, Structural Evaluation, Thermal Evaluation, Containment, Shielding Evaluation, and Criticality Evaluation sections of the application. Results of the Operating Procedures review are considered in the review of Acceptance Tests and Maintenance Program. An example of the information flow for the review of the operating procedures is shown in Figure 7-1.

Additional guidance on operating procedures is provided in NUREG/CR-4775.

7.5.1 Package Loading

7.5.1.1 Preparation for Loading

Review the procedures for preparing the package for loading. At a minimum, the procedures should:

- Specify that the package should be loaded and closed in accordance with written procedures

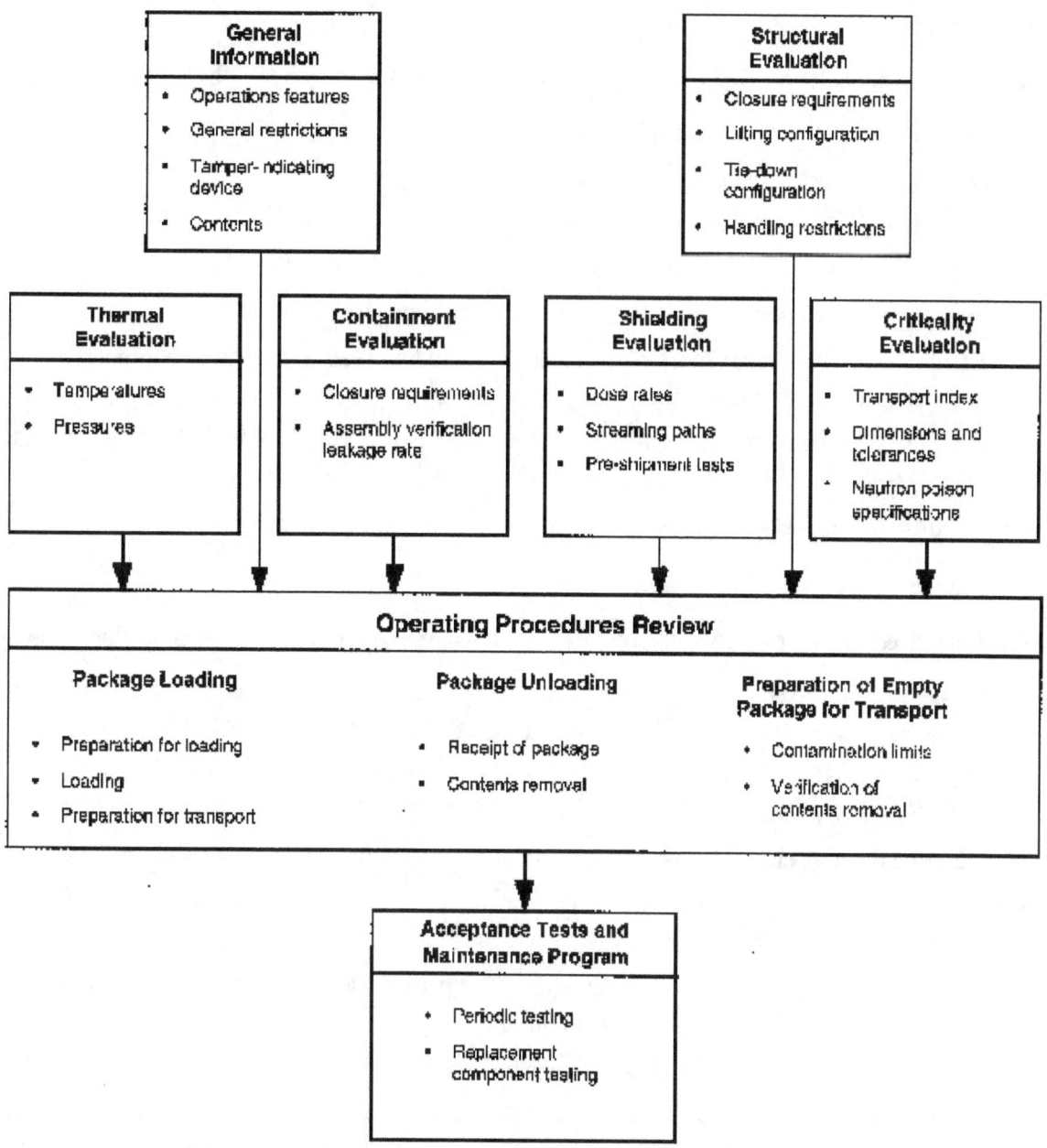

Figure 7-1 Information Flow for the Operating Procedures Review

- Ensure that the contents are authorized in the certificate of compliance
- Ensure that the use of the package complies with the conditions of approval in the certificate of compliance, including verification that required maintenance has been performed
- Verify that the package is in unimpaired physical condition
- Describe any special controls and precautions for handling.

7.5.1.2 Loading of Contents

Review the procedures for loading the contents. At a minimum, the procedures should:

- Identify any special handling equipment needed
- Describe any special controls and precautions for loading
- Indicate the method of loading the contents
- Ensure that any required moderator or neutron absorber is present and in proper condition
- Describe the method to remove water from the package, as appropriate
- Ensure that each closure device of the package, including seals and gaskets, is properly installed, secured, and free of defects
- Verify that the bolt torques described in the procedures are consistent with those shown on the drawings
- Confirm that the package has been loaded and closed appropriately.

7.5.1.3 Preparation for Transport

Review the procedures for preparing the package for transport. At a minimum, the procedures should:

- Ensure that non-fixed (removable) radioactive contamination on external surfaces is as low as reasonably achievable, and within the limits specified in 49 CFR 173.443
- Describe the radiation survey to confirm that the allowable external radiation levels specified in §71.47 are not exceeded
- Describe the temperature survey to verify that limits specified in §71.43(g) are not exceeded
- Specify the assembly verification leakage rate and ensure package closures are leak tested in accordance with ANSI N14.5
- Ensure that any system for containing liquid is properly sealed and has adequate space or other specified provision for expansion of the liquid
- Verify that any pressure relief device is operable and set
- Ensure that any structural component that could be used for lifting or tie-down during transport is rendered inoperable for those purposes unless it meets the design requirements of §71.45
- Ensure that the tamper-indicating device is installed
- Describe, for a fissile material shipment, any special controls and precautions for transport, loading, unloading, and handling and any appropriate actions in case of an accident or delay which should be provided to the carrier or consignee
- Identify any special controls which should be provided to the carrier for a package shipped by exclusive use under the provisions of §71.47(b)(1)
- Describe any special instructions which should be provided to the consignee for opening the package

- Insure that the package is properly labeled.

7.5.2 Package Unloading

7.5.2.1 Receipt of Package from Carrier

Review the procedures for receiving the package. At a minimum, the procedures should:

- Describe any special actions to be taken if the tamper indicating device is not intact, or if surface contamination or radiation survey levels are too high
- Identify any special handling equipment needed
- Describe any proposed special controls and precautions for handling and unloading.

7.5.2.2 Removal of Contents

Review the procedures for unloading a package. At a minimum, the procedures should:

- Describe the appropriate method to open the package
- Identify the appropriate method to remove the contents
- Ensure that the contents are completely removed.

7.5.3 Preparation of Empty Package for Transport

Review the procedures for preparing an empty package for transport. At a minimum, the procedures should:

- Verify that the package is empty
- Ensure that external and internal contamination levels meet the requirements of 49 CFR 173.443 and 49 CFR 173.428
- Describe the package closure requirements.

7.5.4 Other Procedures

Confirm that procedures for any special operational controls are included (e.g., route, weather, or shipping time restrictions).

7.5.5 Appendix

Confirm that the appendix includes a list of references, copies of applicable references if not generally available to the reviewer, test results, and other appropriate supplemental information.

7.6 Evaluation Findings

The Safety Evaluation Report should include a finding similar to the following:

Based on review of the statements and representations in the application, the staff concludes that the operating procedures meet the requirements of 10 CFR Part 71 and that these procedures are adequate to assure the package will be operated in a manner consistent with its evaluation for approval.

7.7 References

American National Standards Institute, ANSI N14.5-1997, "American National Standard for Radioactive Materials–Leakage Tests on Packages for Shipment," New York.

U.S. Nuclear Regulatory Commission, "Guide for Preparing Operating Procedures for Shipping Packages," NUREG/CR-4775, July 1988.

8 ACCEPTANCE TESTS AND MAINTENANCE PROGRAM REVIEW

8.1 Review Objective

The objective of this review is to verify that the acceptance tests for the packaging meet the requirements of 10 CFR Part 71 and that the maintenance program is adequate to assure packaging performance during its service life.

8.2 Areas of Review

The description of the acceptance tests and maintenance program should be reviewed. The review should include:

8.2.1 Acceptance Tests

- Visual Inspections and Measurements
- Weld Examinations
- Structural and Pressure Tests
- Leakage Tests
- Component and Material Tests
- Shielding Tests
- Thermal Tests

8.2.2 Maintenance Program

- Structural and Pressure Tests
- Leakage Tests
- Component and Material Tests
- Thermal Tests
- Miscellaneous Tests

8.2.3 Appendix

8.3 Regulatory Requirements

Regulatory requirements of 10 CFR Part 71 applicable to the review of the acceptance tests and maintenance program are as follows:

8.3.1 Acceptance Tests

- The application must identify codes, standards, and provisions of the quality assurance program used for the acceptance testing of the packaging. [§71.31(c), §71.37(b)]

- Before first use, the fabrication of each packaging must be verified to be in accordance with the approved design. [§71.85(c)]

- Before first use, each packaging must be inspected for cracks, pinholes, uncontrolled voids, or other defects that could significantly reduce its effectiveness. [§71.85(a)]

- Before first use, if the maximum normal operating pressure of a package exceeds 35 kPa (5 psi) gauge, the containment system of each packaging must be tested at an internal pressure at least 50% higher than maximum normal operating pressure to verify its ability to maintain structural integrity at that pressure. [§71.85(b)]

- Before first use, each packaging must be conspicuously and durably marked with its model number, serial number, gross weight, and a package identification number assigned by the NRC. [§71.85(c)]

- The licensee must perform any tests deemed appropriate by the NRC. [§71.93(b)]

8.3.2 Maintenance Program

- The application must identify codes, standards, and provisions of the quality assurance program used for the maintenance program for the packaging. [§71.31(c), §71.37(b)]

- The packaging must be maintained in unimpaired physical condition except for superficial defects such as marks or dents. [§71.87(b)]

- The presence of any moderator or neutron absorber, if required, in a fissile material package must be verified prior to each shipment. [§71.87(g)]

- The licensee must perform any tests deemed appropriate by the NRC. [§71.93(b)]

8.4 Acceptance Criteria

- The acceptance tests and maintenance program must meet the regulatory requirements of Section 8.3.

- Before first use, each packaging must be subject to appropriate acceptance tests to verify that it has been fabricated in accordance with its approved design and that its performance will meet the regulatory requirements of 10 CFR Part 71.

- The maintenance program must be adequate to assure that the package will perform as intended throughout its service life.

8.5 Review Procedures

The review should ensure that appropriate acceptance tests and maintenance program are specified for the package.

The review of the acceptance tests and maintenance program is based in part on the descriptions and evaluations presented in previous sections of the application. An example of the information flow for this review is shown in Figure 8-1.

The commitments specified in the Acceptance Tests and Maintenance Program section of the application are typically included in the certificate of compliance as conditions of the package approval.

8.5.1 Acceptance Tests

Verify that the following tests, as applicable, are to be performed prior to the first use of each packaging. Information presented on each test should include, as a minimum, a description of the test and its acceptance criteria. Applicable sections of the quality assurance program and procedures may be referenced.

Each package must be fabricated in accordance with the drawings listed in the certificate of compliance.

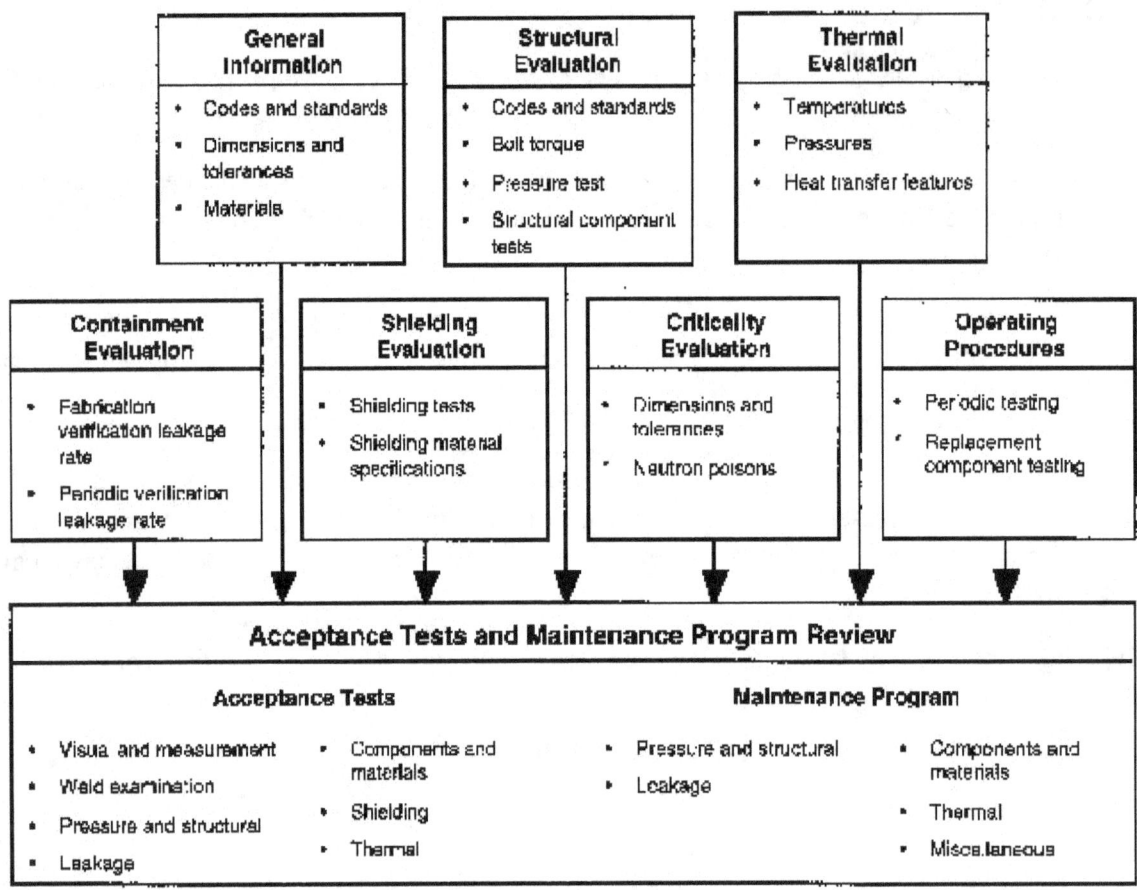

Figure 8-1 Flow Information for the Acceptance Tests and Maintenance Program Review

Additional guidance on acceptance tests is provided in NUREG/CR-3854.

8.5.1.1 Visual Inspections and Measurement

Ensure that inspections are performed to verify that the packaging has been fabricated and assembled in accordance with the drawings. Dimensions and tolerances specified on the drawings should be confirmed by measurement.

8.5.1.2 Weld Examinations

Verify that welding examinations are performed to verify fabrication in accordance with the drawings, codes, and standards specified in the application. Location, type, and size of the welds should be confirmed by measurement. Other specifications for weld performance, nondestructive examination, and acceptance should be verified as appropriate.

Additional guidance on welding criteria is provided in NUREG/CR-3019.

8.5.1.3 Structural and Pressure Tests

Verify that the structural or pressure tests are identified and described. Such tests should comply with §71.85(b), as well as applicable codes or standards specified in the application.

8.5.1.4 Leakage Tests

Verify that the containment system of the packaging will be subjected to the fabrication leakage test specified in ANSI N14.5. Verify that all closures, including drains and vents, are leak- tested. The acceptable leakage criterion should be consistent with that identified in the Containment section.

8.5.1.5 Component and Material Tests

Confirm that appropriate tests and acceptance criteria are specified for components that affect package performance. Examples of such components include seals, gaskets, valves, fluid transport systems, and rupture disks or other pressure-relief devices. Components should be tested to meet the performance specifications shown on the engineering drawing of the package. When tests adversely affect the continued performance of a component, applicable quality assurance procedures should be described to justify that the tested component is equivalent to the component that will be used in the packaging.

Verify that appropriate tests and acceptance criteria are specified for packaging materials. Tests for neutron absorbers (e.g., boron, gadolinia) and insulating materials (e.g., foams, fiberboard) should assure that minimum specifications for density and isotopic content are achieved. Materials should be tested to meet the performance specifications shown on the engineering drawings.

8.5.1.6 Shielding Tests

Ensure that appropriate shielding tests are specified for both neutron and gamma radiation. The tests and acceptance criteria should be sufficient to assure that no voids or streaming paths exist in the shielding.

8.5.1.7 Thermal Tests

Verify that appropriate tests are specified to demonstrate the heat transfer capability of the packaging. These tests should confirm that the heat transfer performance determined in the evaluation is achieved in the fabrication process.

8.5.2 Maintenance Program

Confirm that the maintenance program is adequate to assure that packaging effectiveness is maintained throughout its service life. Maintenance tests and inspections should be described with schedules for each test or replacement of parts and criteria for minor refurbishment and replacement of parts, as applicable.

8.5.2.1 Structural and Pressure Tests

Verify that any periodic structural or pressure tests are identified and described. Such tests would generally be applicable to codes, standards, or other procedures specified in the application.

8.5.2.2 Leakage Tests

Confirm that the containment system of the packaging will be subjected to the periodic and maintenance leakage tests specified in ANSI N14.5. The acceptable leakage criterion should be consistent with that identified in the Containment section. Elastomeric seals should be replaced and leak tested within the 12-month period prior to shipment. Metallic seals are generally replaced prior to each shipment.

8.5.2.3 Component and Material Tests

Verify that periodic tests and replacement schedules for components are described as appropriate.

Confirm that the application identifies any process that could result in deterioration of packaging materials, including loss of neutron absorbers, reduction in hydrogen content of shields, and density changes of insulating materials. Appropriate tests and their acceptance criteria to ensure packaging effectiveness for each shipment should be specified.

8.5.2.4 Thermal Tests

Verify that periodic tests to assure the heat transfer capability during the service life of the packaging are described. Tests similar to the acceptance tests discussed in Section 8.5.1.7 may be applicable. The typical interval for periodic thermal tests is five years.

8.5.2.5 Miscellaneous Tests

Confirm that any additional tests that should be performed periodically on the package or its components are described.

8.5.3 Appendix

Confirm that the appendix includes a list of references, copies of applicable references if not generally available to the reviewer, and other appropriate supplemental information.

8.6 Evaluation Findings

The Safety Evaluation Report should include a finding similar to the following:

> Based on review of the statements and representations in the application, the staff concludes that the acceptance tests for the packaging meet the requirements of 10 CFR Part 71 and that the maintenance program is adequate to assure packaging performance during its service life.

8.7 References

American National Standards Institute, ANSI N14.5-1997, "American National Standard for Radioactive Materials–Leakage Tests on Packages for Shipment," New York.

U.S. Nuclear Regulatory Commission, "Fabrication Criteria for Shipping Containers," NUREG/CR-3854, March 1985.

U.S. Nuclear Regulatory Commission, "Welding Criteria for Use in the Fabrication of Radioactive Material Shipping Containers, NUREG/CR-3019, March 1984.

APPENDIX A1:

RADIOGRAPHY PACKAGES

A1.1 Package Type

A1.1.1 Purpose of Package

These packages include radiographic exposure devices and radiographic source changers. The purpose of an exposure device is to transport a Type B quantity of special form radioactive material for use as a radiographic gamma source. The purpose of the source changer device is to transport a radiographic gamma source to and from an exposure device and to exchange radiographic sources with that exposure device.

A1.1.2 Description of a Typical Package

A typical packaging used as an exposure device consists of a lead or depleted uranium shield inside a welded steel or titanium housing. The shield includes a metallic S-shaped tube that houses the source during transport and allows movement of the source into position for radiography. The shield may be fixed in position by retention cups welded to end plates of the housing and by foam between the shield and the housing.

The source is attached to the end of a short metallic cable, or pigtail. A securing lock mechanism is installed at one end of the housing to maintain the source in a fixed position during transport. A safety plug assembly installed at the other end of the S-tube provides a redundant mechanism to prevent movement of the source toward an outlet.

The content of a package used as an exposure device is one radiographic gamma source (^{60}Co or ^{192}Ir) in Type B special form.

The package is typically hand-carried by one person using a handle attached to the housing, although it is sometimes mounted on wheels.

A typical packaging used as a radiographic source changer is similar to that used as an exposure device. A source changer may contain multiple sources, typically housed in U-shaped tubes. In addition to its function as a transportation package, a source changer is used to move sources either from or to an exposure device. Although the remainder of this appendix specifically addresses exposure devices, the review of a source changer is similar.

A sketch of a typical radiographic exposure device is presented in Figure A1-1.

A1.2 Package Safety

A1.2.1 Safety Functions

The principal safety function of these packages is to retain the radiographic source and to provide gamma shielding. Containment is provided primarily by the special form source itself. These packages do not contain fissile material.

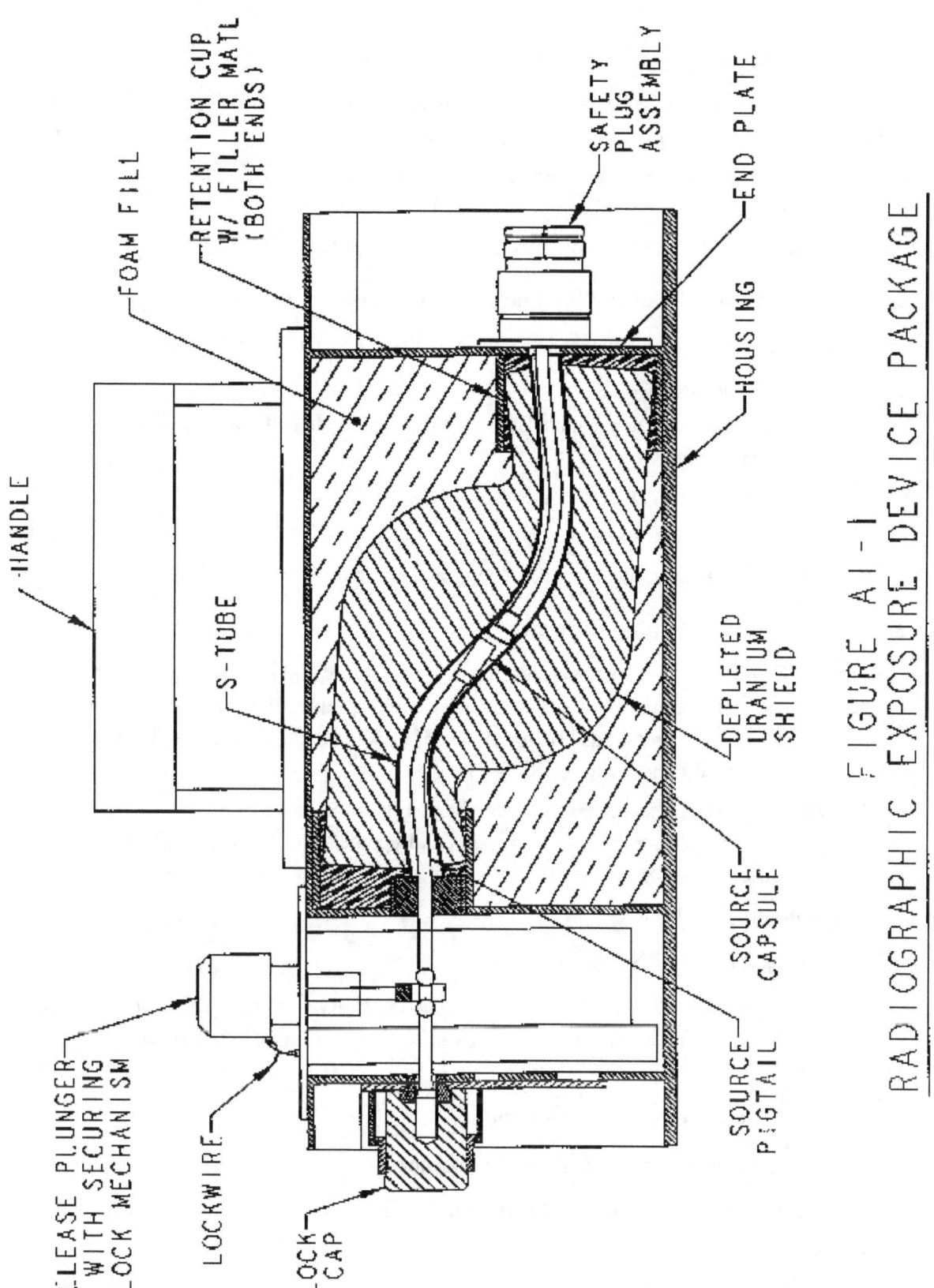

HANDLE

FOAM FILL

RETENTION CUP W/ FILLER MATL (BOTH ENDS)

S-TUBE

SAFETY PLUG ASSEMBLY

END PLATE

HOUSING

DEPLETED URANIUM SHIELD

SOURCE CAPSULE

SOURCE PIGTAIL

RELEASE PLUNGER WITH SECURING LOCK MECHANISM

LOCKWIRE

LOCK CAP

FIGURE A1-1

RADIOGRAPHIC EXPOSURE DEVICE PACKAGE

A1.2.2 Safety Features

- A lead or depleted uranium shield provides gamma shielding.

- A securing lock mechanism positions the source pigtail within the S-tube in the shield during transport to prevent high radiation fields and radiation streaming.

- A safety plug assembly at the opposite end of the tube provides a redundant mechanism to prevent movement of the source.

- The housing, foam, and other structural materials protect the shield and S-tube from damage.

A1.2.3 Typical Areas of Review for Package Drawings

- Housing features, including dimensions, material, thickness, and welds

- Foam material and density

- Shield dimensions and material, including supplemental shielding, its weight, dimensions, and method of attachment

- Material, wall thickness and curvature of S- or U-tube

- Lock mechanism specifications

- Other structural features, including bolts, pins, and retention cups, as applicable.

A1.2.4 Typical Areas of Safety Review

- The general information review verifies that the contents are restricted to special form and that the source nuclide and maximum allowable activity are specified. Specification of content activity is typically expressed as "Bq (output)" or "Ci (output)" to denote that the activity is determined in accordance with ANSI N432-1980.

- The structural and thermal reviews evaluate the ability of the shield to perform its intended function under normal conditions of transport and hypothetical accident conditions. These reviews address:

    - Damage to the shielding

    - Misalignment of the S-tube

    - Damage to the S-tube resulting in exposure of the depleted uranium shield and possible oxidation of the uranium or eutectic reaction between the uranium and other package components

    - Damage to the securing lock mechanism

    - Movement of the source relative to the shielding.

- The shielding review evaluates the ability of the package to satisfy the maximum allowable external radiation levels under normal conditions of transport and hypothetical accident conditions. Shielding requirements are often demonstrated by measuring the dose rates from a gamma test source and scaling them according to the maximum allowed activity of the contents. Key issues include:

- Ensuring that the locations of the maximum radiation levels on the surface of the package, including near the ends of the S-shaped source tube, and at one meter from the surface have been identified

- Determining that the size (active depth and diameter) of the detector is appropriate for providing dose rate measurements at the regulatory locations (because of the small size of the package, corrections may be needed to account for the size of the detector probe volume) (ANSI N43.9-1991)

- Examining the design of the source assembly and securing lock mechanism, including pigtail and locking balls. A small movement in source position can result in a significant increase in external radiation levels

- Verifying that no significant increase in radiation occurs as a result of the tests for normal conditions of transport

- Confirming that the radiation levels under normal conditions of transport and hypothetical accident conditions are satisfied.

- The review of operating procedures confirms that the source is securely locked in position before shipment. This review also evaluates procedures to verify by physical means that the source has been removed before shipment of an "empty" package. Because of shielding effectiveness and radiation from uranium shielding itself, verification by radiation measurements alone may not be sufficient. The procedure should be capable of detecting remaining sources if the pigtail is clipped off.

- The review of acceptance tests and maintenance program verifies that appropriate fabrication and periodic verification tests are performed to demonstrate effectiveness of the shielding. The review also verifies that appropriate inspections are performed to monitor any wearing of the S-tube.

Several NRC Information Notices (85-07, 87-47, 88-18, 88-33, 90-24, 90-35, 90-82) provide additional detail on safety issues relevant to the transport of radiography packages.

References

American National Standards Institute, ANSI N43.9-1991, "American National Standard for Gamma Radiography—Specifications for Design and Testing of Apparatus," New York.

National Bureau of Standards, "Radiological Safety for the Design and Construction of Apparatus for Gamma Radiography," ANSI N432-1980, Washington, DC.

U.S. Nuclear Regulatory Commission, "Contaminated Radiography Source Shipments," NMSS Information Notice 85-07, January 29, 1985.

U.S. Nuclear Regulatory Commission, "Malfunction of Lockbox on Radiography Device," NMSS Information Notice 88-18, April 25, 1988.

U.S. Nuclear Regulatory Commission, "Recent Problems Involving the Model SPEC 2-T Radiographic Exposure Device," NMSS Information Notice 88-33, May 27, 1988.

U.S. Nuclear Regulatory Commission, "Requirements for Use of Nuclear Regulatory Commission- (NRC-) Approved Transport Packages for Shipment of Type A Quantities of Radioactive Material," NMSS Information Notice 90-82, December 31, 1990.

U.S. Nuclear Regulatory Commission, "Transportation of Model SPEC 2-T Radiographic Exposure Device," NMSS Information Notice 90-24, April 10, 1990.

U.S. Nuclear Regulatory Commission, "Transportation of Radiography Devices SSINS No. 6835," NMSS Information Notice 87-47, October 5, 1987.

U.S. Nuclear Regulatory Commission, "Transportation of Type A Quantities of Radioactive Materials," NMSS Information Notice 90-35, May 24, 1990.

APPENDIX A2:

TYPE B WASTE PACKAGES

A2.1 Package Type

A2.1.1 Purpose of Package

The purpose of this type of package is to transport a Type B quantity of dry, radioactive, irradiated, and contaminated solid materials.

A2.1.2 Description of a Typical Package

A typical packaging consists of a steel-encased, lead-shielded cylinder with impact limiters attached at both ends. The packaging may be protected by a thermal shield, consisting of a thin metal shell separated from the lead-filled cylinder by a wire wrap. Closure is provided by a bolted steel lid, which may also include lead shielding. Two concentric O-rings are installed in grooves typically on the underside of the lid. The lid includes a leak test port between the O-rings and sometimes a vent port. The bottom of the packaging contains a sealed drain port.

A typical packaging may be sized to transport ion-exchange resins, process solids, or irradiated hardware, such as control rod blades. It is approximately 3.3 m (about 11 ft.) in length and 1.3 m (about 4 ft.) in diameter (without impact limiters) and can weigh as much as 35 tons (without contents). The packaging generally has two or four trunnions near the top for lifting, and two near the bottom for rotation.

The contents of the package consist of a Type B quantity of dry, radioactive, irradiated and contaminated solid materials, generally within a secondary container. The maximum content weight may approach five tons, including shoring. The radioactive contents typically include waste containing mixed fission products and activation products. The fissile material content of these packages is limited to that permitted by the 10 CFR Part 71 general license provisions for fissile material packages, or exempt quantities.

A sketch of a typical Type B waste package is presented in Figure A2-1.

A2.2 Package Safety

A2.2.1 Safety Functions

The principal safety function of the package is to provide gamma shielding and containment.

A2.2.2 Safety Features

- The lead shield provides gamma shielding. The neutron source is not significant.
- The inner vessel provides containment of the radioactive material. Although secondary containers are often used, they do not provide a containment function.

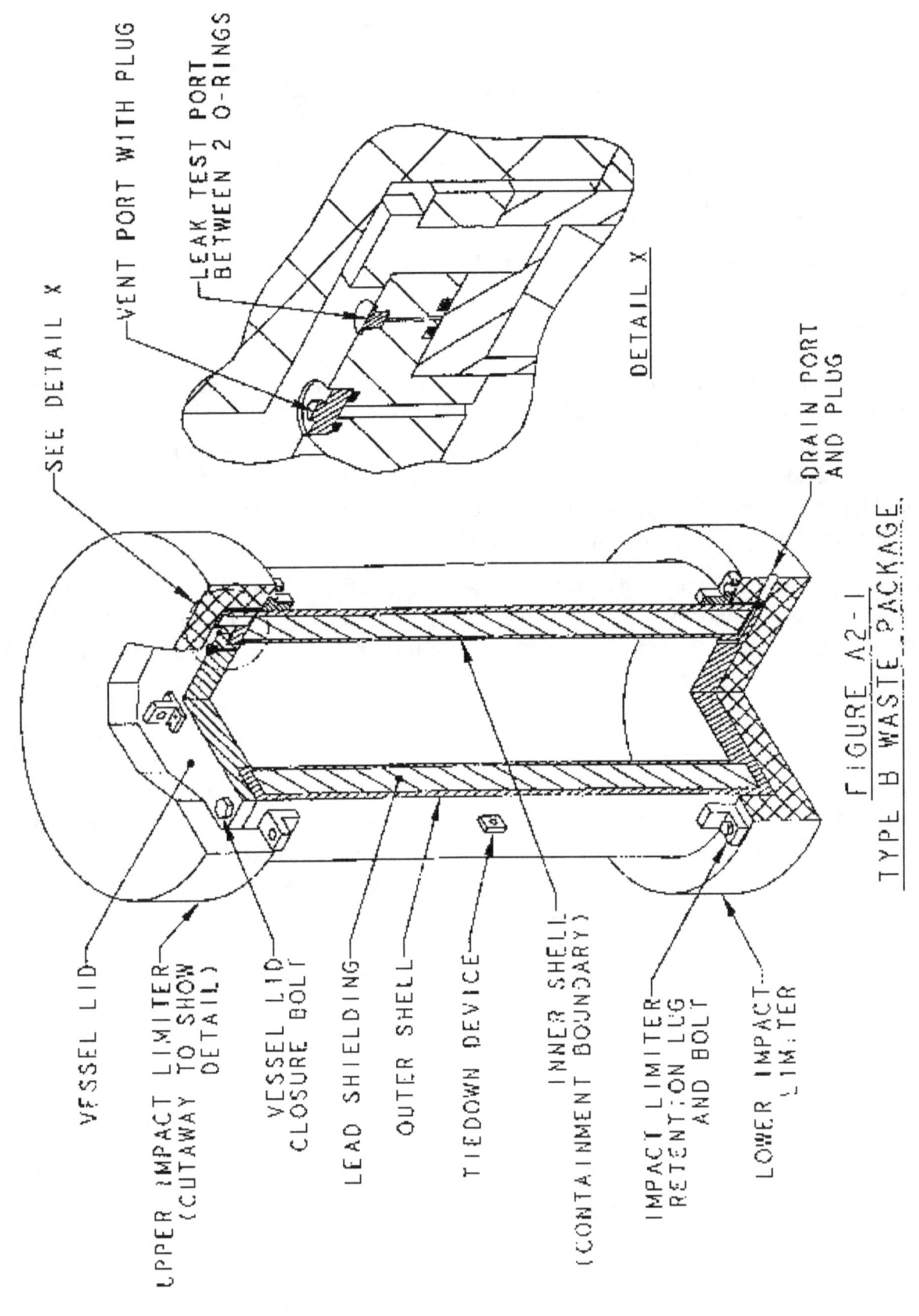

VENT PORT WITH PLUG

LEAK TEST PORT
BETWEEN 2 O-RINGS

SEE DETAIL X

DETAIL X

DRAIN PORT
AND PLUG

FIGURE A2-1

TYPE B WASTE PACKAGE

VESSEL LID

UPPER IMPACT LIMITER
(CUTAWAY TO SHOW
DETAIL)

VESSEL LID
CLOSURE BOLT

LEAD SHIELDING

OUTER SHELL

TIEDOWN DEVICE

INNER SHELL
(CONTAINMENT BOUNDARY)

IMPACT LIMITER
RETENTION LUG
AND BOLT

LOWER IMPACT
LIMITER

A2.2.3 Typical Areas of Review for Package Drawings

- Containment vessel body
  - Materials of construction
  - Dimensions and tolerances of structural shell and shielding material
  - Fabrication codes or standards
  - Weld specifications, including codes or standards for nondestructive examination
  - Thermal shield, if applicable.

- Containment vessel closures
  - Lid materials, and their dimensions and tolerances
  - Bolt specifications, including number, size, minimum thread engagement, and torque
  - Seal material, size, and compression specifications
  - Seal groove dimensions
  - Vent, drain, and leak-test ports, including closure methods.

- Impact limiters
  - Materials of construction and dimensions
  - Foam or wood specifications, including density
  - Method of attachment.

A2.2.4 Typical Areas of Safety Review

- The general information review identifies the allowable contents, including water and other materials that could produce combustible gases.

- The structural and thermal reviews evaluate the performance of the containment system during both normal conditions of transport and hypothetical accident conditions. Primary emphasis is on the structural and thermal effects at the closure regions (lid and ports), including O-rings, plugs, and bolts, under hypothetical accident conditions.

- The structural and thermal reviews also verify the effects of the hypothetical accident conditions tests on the lead shielding and thermal shield (if applicable).

- The thermal review confirms the maximum temperature and pressure in the containment vessel under normal conditions of transport and hypothetical accident conditions.

- The containment review verifies that the package closures (lid, vent port, drain port) meet 10 CFR Part 71 containment criteria using the methods in ANSI N14.5 for both normal conditions of transport and hypothetical accident conditions. A typical maximum allowable leakage rate is approximately 10^{-5} ref cc/s. The review also confirms that combustible-gas generation meets the criteria discussed in Section 4 of this document.

- The shielding review confirms that the package meets the allowable radiation levels during both normal conditions of transport and hypothetical accident conditions. The review should also confirm that the lead shielding does not melt under the hypothetical accident conditions.

- The criticality review verifies that the package contains either no fissile material, an exempt quantity of fissile material, or a fissile material quantity allowed under the general license provisions of 10 CFR Part 71. For packages with fissile content limited to quantities authorized by general license, the review also should confirm that the correct criticality transport index is specified.

- The review of operating procedures verifies that the bolts are properly torqued and that all penetrations of the containment vessel are properly leak tested prior to shipment. The review also addresses procedures that assure the contents are dry.

- The review of acceptance tests and maintenance program confirms that the appropriate leakage tests are performed for fabrication and periodic verification during the service life of the package. The review also ensures that appropriate acceptance testing of the lead shield and thermal performance is described and that the thermal performance of the packaging is maintained during the service life.

References

American National Standards Institute, ANSI N14.5-1997, "American National Standard for Radioactive Materials–Leakage Tests on Packages for Shipment," New York.

APPENDIX A3:

UNIRRADIATED FUEL PACKAGES

A3.1 Package Type

A3.1.1 Purpose of Package

The purpose of this type of package is to transport unirradiated fuel assemblies and individual fuel rods. These packages are also referred to as "fresh fuel packages."

This appendix addresses only those packages in which the contents are limited to a Type A quantity of fissile material. For entire assemblies, this is typically achieved by restricting the enrichment. For individual fuel rods, a combination of enrichment and mass limits may be specified.

A3.1.2 Description of a Typical Package

A typical packaging consists of a metal outer shell, closed with bolts and a weather-tight gasket. An internal steel strongback, shock-mounted to the outer shell, supports one or two fuel assemblies, which are fixed in position on the strongback by clamps, separator blocks, and end support plates. Depending on the type of fuel, neutron poisons are sometimes used to reduce reactivity. If the package is used to transport individual fuel rods, a separate inner container is often employed.

The contents of the package are unirradiated uranium in fuel assemblies or individual fuel rods. Because the majority of these packages are for commercial reactor fuel, the uranium is typically in the form of Zircaloy-clad uranium dioxide pellets.

Sketches of the typical package described above are presented in Figures A3-1 and A3-2.

A3.1.3 Alternative Package Design

An alternative design for a fresh fuel package is shown in Figure A3-3. In this design, the fuel assemblies are fixed in position by two steel channels, mounted by angle irons or a similar bracing structure to a thin-walled inner metal container. This inner container is in turn surrounded by a honeycomb material and enclosed in a wooden outer container. Foam cushioning material is also generally used to cushion the fuel assemblies and may be used between the inner and outer container.

A3.2 Package Safety

A3.2.1 Safety Functions

The principal function of the package is to provide criticality control. The metal outer shell of the packaging retains the assemblies within a fixed geometry relative to other such packages in an array and provides impact and thermal protection. Shielding requirements are not significant because of the low radioactivity of unirradiated fuel.

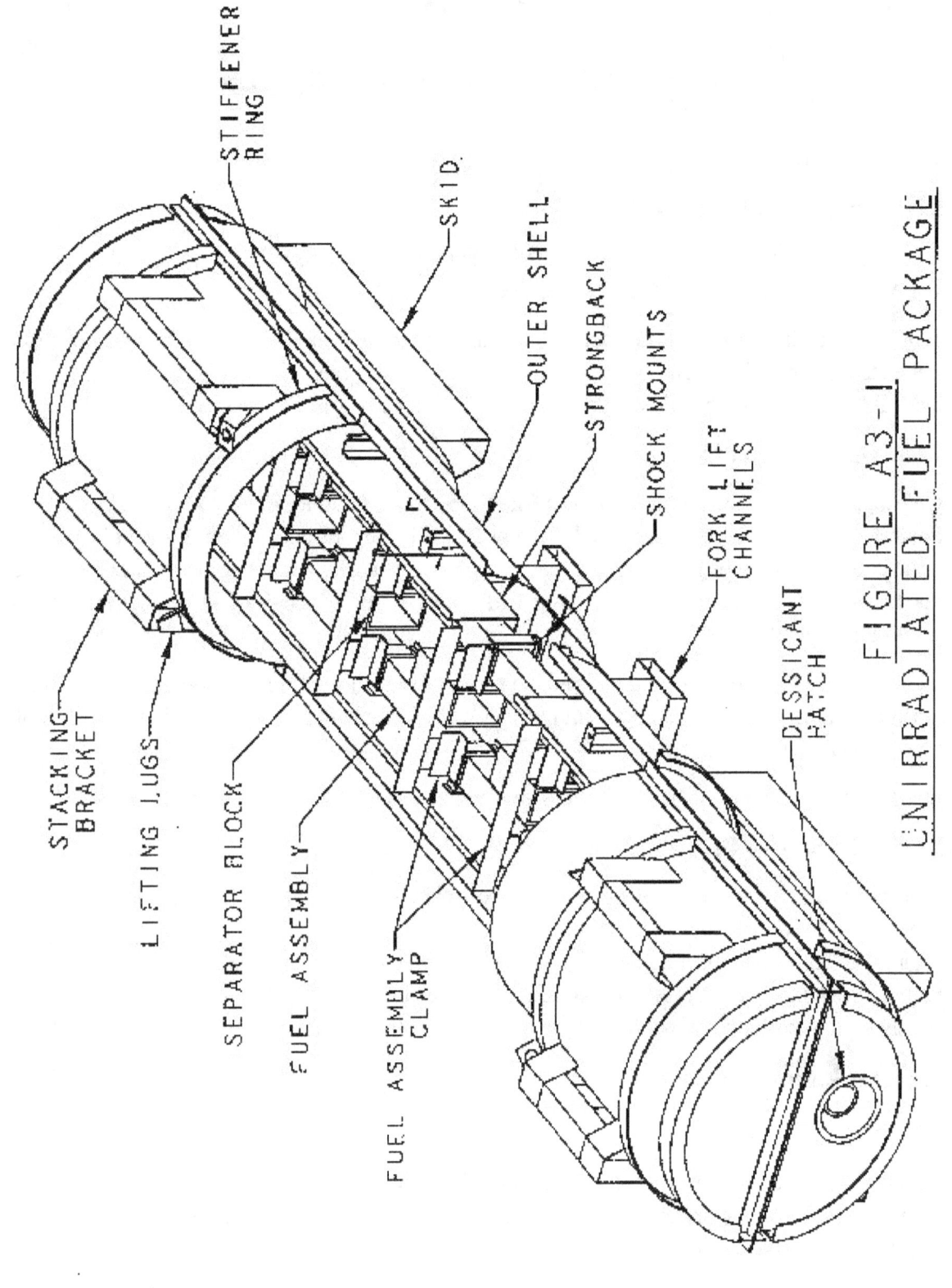

STIFFENER RING

STACKING BRACKET

SKID

OUTER SHELL

LIFTING LUGS

STRONGBACK

SEPARATOR BLOCK

SHOCK MOUNTS

FUEL ASSEMBLY

FORK LIFT CHANNELS

FUEL ASSEMBLY CLAMP

DESSICANT HATCH

FIGURE A3-1

UNIRRADIATED FUEL PACKAGE

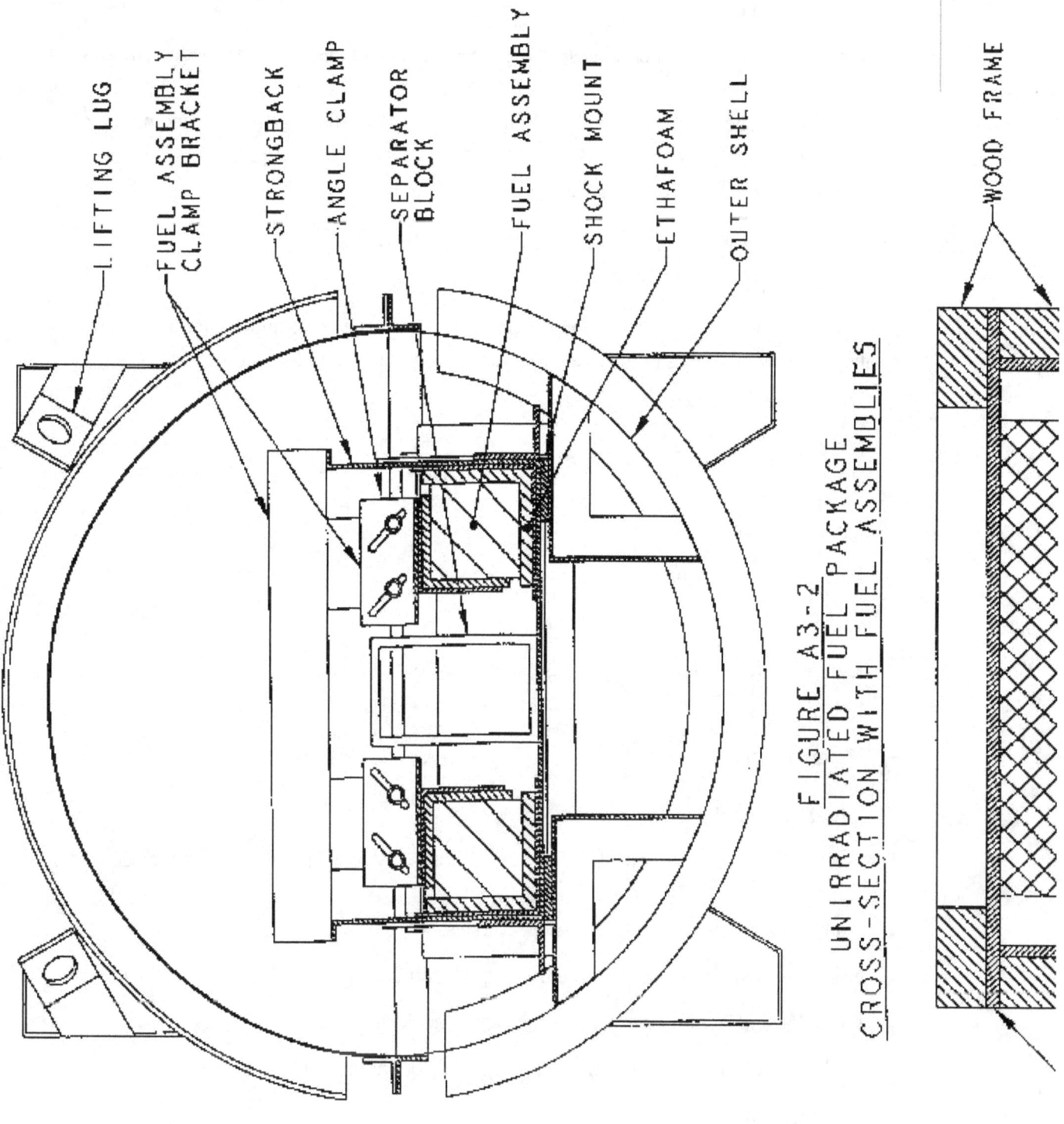

LIFTING LUG

FUEL ASSEMBLY CLAMP BRACKET

STRONGBACK

ANGLE CLAMP

SEPARATOR BLOCK

FUEL ASSEMBLY

SHOCK MOUNT

ETHAFOAM

OUTER SHELL

WOOD FRAME

FIGURE A3-2
UNIRRADIATED FUEL PACKAGE
CROSS-SECTION WITH FUEL ASSEMBLIES

A3.2.2 Safety Features

- A strongback with end support plates, clamps, and separators maintains the fuel assemblies in a fixed position relative to each other and to any neutron poisons.

- The metal outer shell of the packaging retains and protects the fuel assemblies, and may provide a minimum spacing between assemblies in an array of packages.

- Neutron poisons, if present, reduce reactivity.

A3.2.3 Typical Areas of Review for Package Drawings

- Outer shell dimensions
- Structural components (e.g., strongback, support plates, fuel clamps and separators) that fix the position of fuel assemblies or relative position between fuel assemblies and poisons
  - Dimensions and materials
  - Methods of attachment.

- Neutron poisons
  - Dimensions and tolerances
  - Minimum poison content
  - Location and method of attachment.

- Moderating materials, including plastics, wood, and foam
  - Location
  - Material properties.

A3.2.4 Typical Areas of Safety Review

- The general information review identifies the fuel assembly designs authorized in the package, including:
  - Number of and arrangement of fuel assemblies
  - Number, pitch, and position of fuel rods, guide tubes, and channels
  - Overall assembly dimensions, including active fuel length
  - Authorization or restrictions on missing fuel rods or partial-length rods
  - Maximum enrichment
  - Pellet dimensions and tolerances
  - Minimum cladding thickness
  - Fuel-clad gap

- Type, location, and concentration of burnable poisons

- Type, location, and quantity of plastics, such as polyethylene, within the fuel assemblies.

- The structural review addresses possible damage to the outer shell, strongback, fuel assembly, neutron poisons (if present), clamps, separators, and end support plates to ensure that the fuel assemblies and neutron poisons are maintained in a fixed position relative to each other under hypothetical accident conditions.

- The structural review also confirms the minimum spacing between fuel assemblies in different packages in an array under hypothetical accident conditions. Spacing can be affected by separation of the strongback from its shock mounts, failure of the shock mounts or fuel assembly clamps, and deformation of the outer shell of the package.

- The thermal review evaluates the effect of the fire on neutron poisons, plastic sheeting, wood, or other temperature-sensitive materials under hypothetical accident conditions.

- The criticality review addresses both normal conditions of transport and hypothetical accident conditions. Key areas for this review include:

  - The number of packages in the array and the array configuration (pitch, orientation of packages, etc.). Because of movement of the strongback within the package and the location of poisons, the arrays might not be symmetrical.

  - Degree of moderation. Structural features, as well as packaging material such as plastic sheeting, are evaluated for the possibility of differential flooding within the package. Plastic sheeting on the fuel assemblies should be open at both ends to preclude differential flooding. Flooding between the fuel pellets and cladding is also considered. Variations in the allowable amount of light-weight packaging material and plastic shims inserted in the fuel assemblies can also affect criticality under normal conditions of transport.

- The review of operating procedures ensures that instructions are provided so that proper clamps, separators, and poisons are selected for the type of fuel assemblies to be shipped and that these items are properly installed prior to shipment. The procedures should also address any other restrictions (e.g., limits on number of shims) considered in the package evaluation.

- The review of the acceptance tests and maintenance program verifies that the neutron poisons, if present, are subject to appropriate tests to verify their concentration and uniformity.

APPENDIX A4:

LOW ENRICHED URANIUM OXIDE PACKAGES

A4.1 Package Type

A4.1.1 Purpose of Package

The purpose of this type of package is to transport pellets and powder of low enriched uranium oxide. These packages are also referred to as "low enriched pellet and powder packages" or "oxide packages."

This appendix addresses only those packages in which the contents are limited to a Type A quantity of fissile material. This is achieved by limiting either the maximum enrichment or a combination of enrichment and mass.

A4.1.2 Description of a Typical Package

A typical packaging consists of an inner steel vessel positioned within an outer steel drum. The outer drum, is typically a 30- or 55-gal. steel drum with a removable head and weather-tight gasket. The head is usually secured by a clamp ring with a closure bolt and a tamperproof seal. Vent holes near the top of the drum, which provide pressure relief under hypothetical accident conditions, are capped or taped during normal conditions of transport to prevent water inleakage.

The inner vessel is typically flanged, with a gasket and a bolted lid. The inner vessel is the containment vessel. It is centered in position inside the outer drum by foam, fiberboard, or similar insulation material. The inner vessel is not a pressure vessel and is not designed to prevent water inleakage under hypothetical accident conditions.

The contents of this package include low enriched uranium pellets, powder, and sometimes scrap, which are placed in plastic bags, metal cans, or cardboard boxes prior to loading into the inner container. Pellets are generally arranged on metal trays. Packages may include plates or liners with neutron poisons within the inner vessel. Spacers may be used within the inner vessel to maintain the position of the contents and to displace moderator in the event of water inleakage.

A sketch of a typical package for pellets or powder of low enriched uranium oxide is presented in Figure A4-1.

A4.2 Package Safety

A4.2.1 Safety Functions

The principal function of the package is to provide criticality control. The inner vessel provides containment to satisfy the requirements for Type A packages. Shielding requirements are not significant because of the low radioactivity of unirradiated uranium oxide.

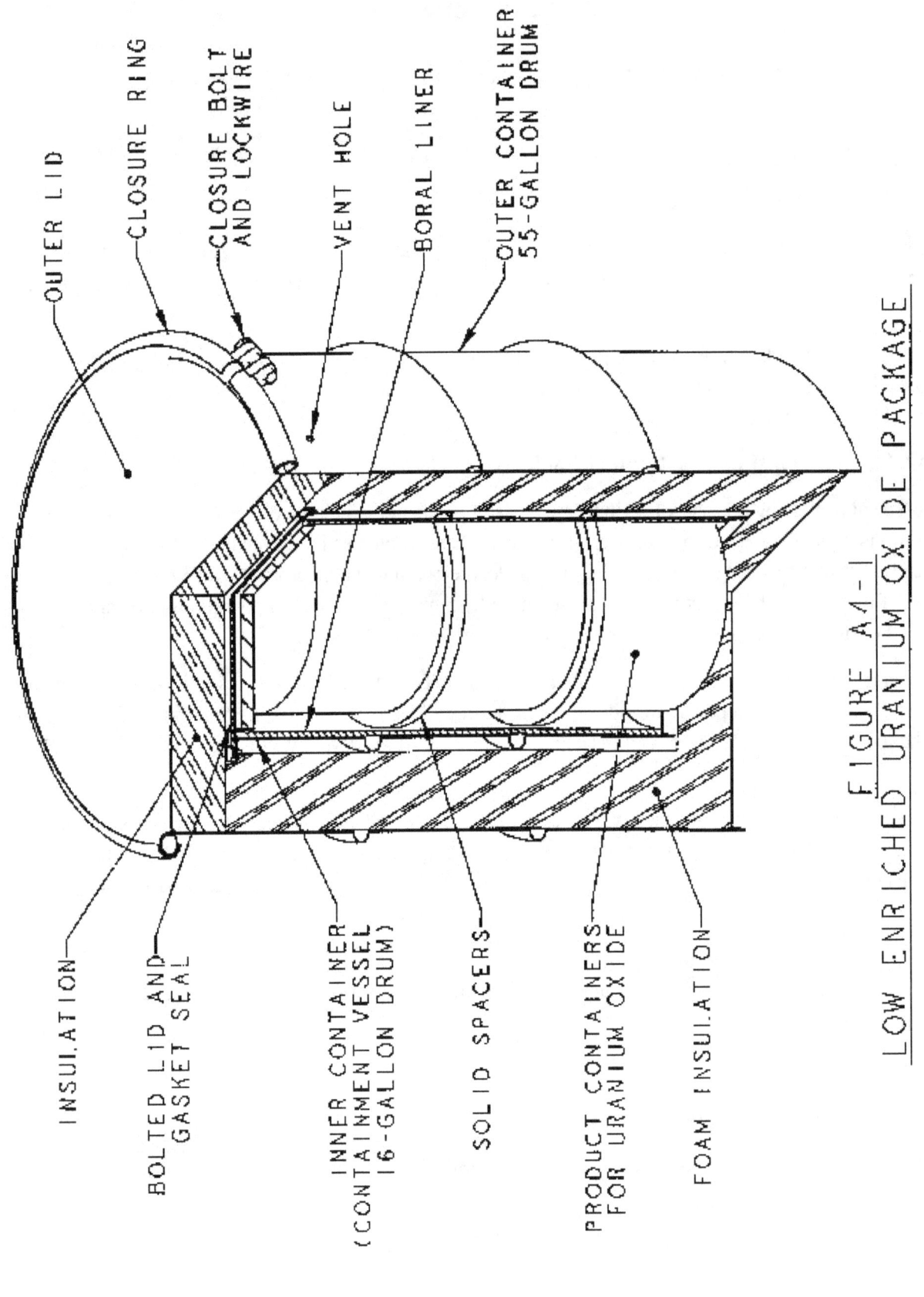

OUTER LID

CLOSURE RING

CLOSURE BOLT
AND LOCKWIRE

VENT HOLE

BORAL LINER

OUTER CONTAINER
55-GALLON DRUM

INSULATION

BOLTED LID AND
GASKET SEAL

INNER CONTAINER
(CONTAINMENT VESSEL
16-GALLON DRUM)

SOLID SPACERS

PRODUCT CONTAINERS
FOR URANIUM OXIDE

FOAM INSULATION

FIGURE A4-1

LOW ENRICHED URANIUM OXIDE PACKAGE

A4.2.2 Safety Features

- The outer metal drum and insulation protect the inner vessel under hypothetical accident conditions, and maintain a minimum spacing between the inner containers of different packagings.
- The inner vessel provides containment and maintains a fixed geometry for criticality control.
- Neutron poisons, if present, reduce reactivity.

A4.2.3 Typical Areas of Review for Package Drawings

- Inner vessel
  - Materials of construction
  - Dimensions and tolerances, including thickness
  - Product containers
  - Spacers, including materials and dimensions
  - Fabrication codes or standards.
- Neutron poisons
  - Isotopes and minimum concentration
  - Dimensions and tolerances
  - Location.
- Insulating material
  - Type
  - Dimensions and tolerances
  - Density.
- Outer drum
  - Material
  - Closure, including use of heavy-duty clamp ring, bolt torque
  - Dimensions.

A4.2.4 Typical Areas of Safety Review

- The structural review evaluates package integrity under drop, puncture and thermal tests. This includes verifying that the lid of the outer drum remains in place and that the inner vessel is not damaged.

- The structural and thermal reviews address the minimum spacing between contents of different packages under hypothetical accident conditions. Damage to outer drum and charring of the insulation may result in closer spacing and more reactivity than that under normal conditions of transport.

- The thermal review also evaluates the effect of fire on neutron poisons and spacers.

- The criticality review addresses in detail both normal conditions of transport and hypothetical accident conditions. Key areas for this review include:

  - The configuration of the contents under normal conditions of transport and hypothetical accident conditions. This includes number, spacing, size, and condition of pellets, distribution of powders, and similar effects. Small changes in dimensions of the inner vessel can result in a significant increase in reactivity.

  - Distribution and degree of moderation. In addition to the moisture content of the pellets or powder, structural features, spacers, and packaging material such as plastic bags or cans are evaluated for the possibility of differential flooding within the package. Variations in the allowable amount of light-weight packaging material are also verified. Loading less than the maximum allowed contents can provide additional volume for water inleakage under hypothetical accident conditions, and therefore partial loads are often more reactive than a fully packed inner vessel.

  - The number of packages considered in the array and the array configuration (e.g., pitch and orientation of packages). Depending on the positioning of contents and the location of poisons, the arrays might not be symmetrical.

  - The degree and location of damage (e.g., drying or charring) to the thermal insulation caused by the fire test.

- The review of operating procedures ensures that instructions are provided so that proper neutron poisons or spacers are selected for the type of contents to be shipped and that the package is properly closed.

- The review of the acceptance tests and maintenance program verifies that the neutron poisons, if present, are subject to appropriate tests to verify their concentration.

APPENDIX A5:

TRANSURANIC WASTE PACKAGES

A5.1 Package Type

A5.1.1 Purpose of Package

The purpose of this type of package is to transport a Type B quantity of contact-handled transuranic waste.

A5.1.2 Description of a Typical Package

A typical packaging consists of a stainless steel inner containment vessel housed inside a stainless steel and polyurethane outer containment assembly.

The outer containment vessel is a right circular cylinder with a flat bottom and domed lid. Its body and dome generally consist of polyurethane foam sandwiched between an inner and outer stainless steel shell. The dome-shaped lid is secured to the body by a locking ring. An elastomeric O-ring is used as the containment seal; a second O-ring allows the seal to be leak-tested. The assembly typically contains a leak-test port and a vent port. Fork pockets are often located at the base of the assembly for lifting and handling the entire package. Separate lifting devices are used for handling the lid only.

The inner containment vessel is a stainless steel shell with domed ends. The closure system consists of two O-rings, a leak-test port, and a vent port, similar to the outer containment vessel. Lifting devices on the inner lid can be used for lifting either the lid itself or an empty inner containment vessel.

The contents of the package consist of contact-handled transuranic waste produced primarily from plutonium production operations. The waste may be packaged within secondary containers. The contents may be limited to restrict the generation of hydrogen or other combustible gases.

Several packages may be secured to a special trailer for transport.

A sketch of a typical transuranic waste package is presented in Figure A5-1.

A5.2 Package Safety

A5.2.1 Safety Functions

The principal safety functions of the package are to provide containment and criticality control.

A5.2.2 Safety Features

- The inner and outer containment vessels provide double containment for the plutonium.

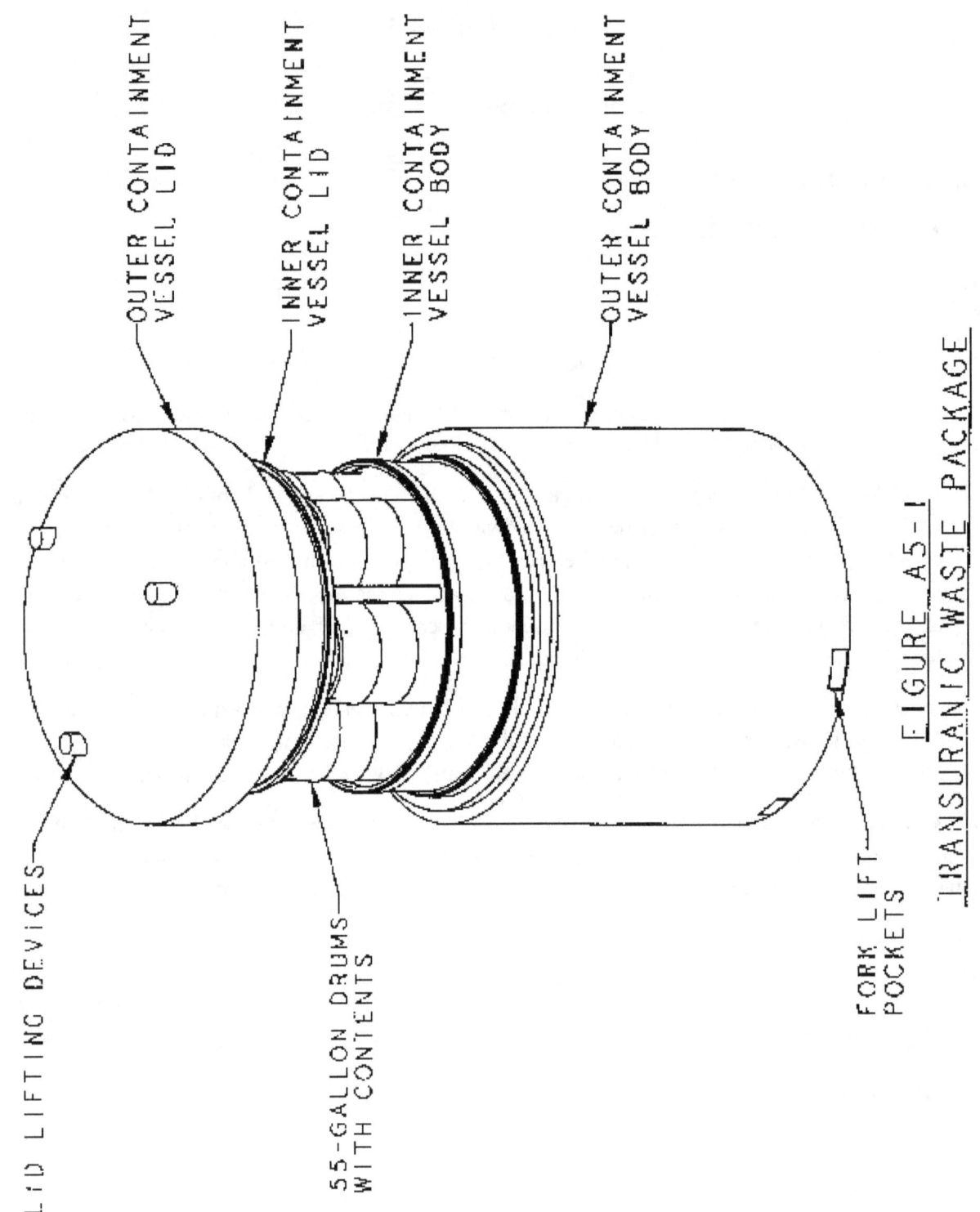

OUTER CONTAINMENT VESSEL LID

INNER CONTAINMENT VESSEL LID

INNER CONTAINMENT VESSEL BODY

OUTER CONTAINMENT VESSEL BODY

LID LIFTING DEVICES

55-GALLON DRUMS WITH CONTENTS

FORK LIFT POCKETS

FIGURE A5-1

TRANSURANIC WASTE PACKAGE

- The steel package and configuration of the secondary containers provide sufficient attenuation and distance from the waste to satisfy the shielding requirements for normal conditions of transport (exclusive use) and hypothetical accident conditions.
- The limit on the allowed mass of fissile material provides criticality control for a single package. The physical size and separation of contents also ensures subcriticality for arrays.

A5.2.3 Typical Areas of Review of Package Drawings

- Containment vessels
  - Materials of construction
  - Dimensions and tolerances
  - Fabrication codes or standards
  - Weld specifications, including codes or standards for nondestructive examination
  - Foam specification and density, as applicable.

- Containment vessel closures
  - Lid materials, and their dimensions and tolerances
  - Closure device design details, such as bolt specifications and torque
  - Seal material, size, and compression specifications
  - Seal groove dimensions
  - Vent and leak-test ports, including closure methods.

A5.2.4 Typical Areas of Safety Review

- The structural and thermal reviews evaluate the ability of the containment vessels to perform their intended functions under normal conditions of transport and hypothetical accident conditions. Primary emphasis is on the structural effects near the O-ring regions (including closure devices) and on the thermal performance of the O-rings.
- The thermal and containment reviews verify that the hydrogen concentration in any confined volume will not exceed 5% (by volume) during a period of one year. Shorter time periods have been approved based on detailed operating procedures to control and track the shipment of packages. The reviews also should ensure that the containment evaluation specifies that the secondary containers are aspirated prior to shipment.
- The containment review verifies that the 10 CFR Part 71 containment criteria are satisfied for both normal conditions of transport and hypothetical accident conditions. With typical contents, the package must remain leaktight, as defined in ANSI N14.5. Each containment vessel must separately meet the 10 CFR Part 71 containment criteria.

- The shielding review evaluates the ability of the package to satisfy the allowed radiation levels during normal conditions of transport and hypothetical accident conditions.

- The criticality review confirms that a single package and array of packages are subcritical during both normal conditions of transport and hypothetical accident conditions.

- The review of operating procedures verifies that any free-standing water is removed from both containment vessels and that they are closed and leak-tested prior to shipment. The review also typically ensures that the secondary containers are aspirated prior to shipment.

- The review of acceptance tests and maintenance program verifies that appropriate fabrication and periodic verification leakage tests are performed.

References

American National Standards Institute, ANSI N14.5-1997, "American National Standard for Radioactive Materials–Leakage Tests on Packages for Shipment," New York.

APPENDIX A6:

LOW ENRICHED URANIUM HEXAFLUORIDE PACKAGES

A6.1 Package Type

A6.1.1 Purpose of Package

The purpose of this type of package is to transport low-enriched solid uranium hexafluoride (UF_6).

A6.1.2 Description of a Typical Package

A typical packaging consists of an inner steel cylinder that acts as a containment vessel, and an outer protective overpack. Unenriched UF_6 may be transported in bare cylinders, without the protective overpack, as authorized in DOT regulations. Protective overpacks are typically required only for the transport of enriched (fissile) UF_6. ANSI N14.1 specifies the design and fabrication of the UF_6 cylinder. ANSI N14.1 and USEC-651 contain information regarding overpacks.

The inner cylinder is carbon steel, with rounded ends and a protective skirt. On one end of the cylinder is a valve for filling and emptying the cylinder; on the other end is a removable plug. The most commonly used commercial cylinders are approximately 0.76 m (30 in.) in diameter, 2.1 m (81 in.) in length, with a capacity of about 2300 kg (2.5 tons) of UF_6. The design and authorized contents are defined in ANSI N14.1.

The protective overpack is generally a double-shell, stainless steel cylinder with cushioning pads on the inner cavity. An energy-absorbing, insulating foam fills the space between the inner and outer shell. The overpack can be separated into two halves to enable easy access to the inner cylinder. Overpacks for the 30-in. cylinders mentioned above are approximately 0.016 m (4 in.) thick.

For the 30-in. cylinder, the UF_6 enrichment must not exceed 5%. The cylinder is filled with liquid UF_6. Because of the volume reduction during cooling and solidification of the UF_6, the final internal pressure is less than one atmosphere in the cylinder.

A sketch of a typical UF_6 package (cylinder and overpack) is presented in Figure A6-1.

A6.2 Package Safety

A6.2.1 Safety Functions

The primary function of the package is to provide containment and moderation control for criticality purposes. Moderation control is required for all commercially used cylinders for fissile UF_6 and must be maintained under normal conditions of transport and hypothetical accident conditions. To assure subcriticality by moderation control, the mass of the contents must be at least 99.5% UF_6.

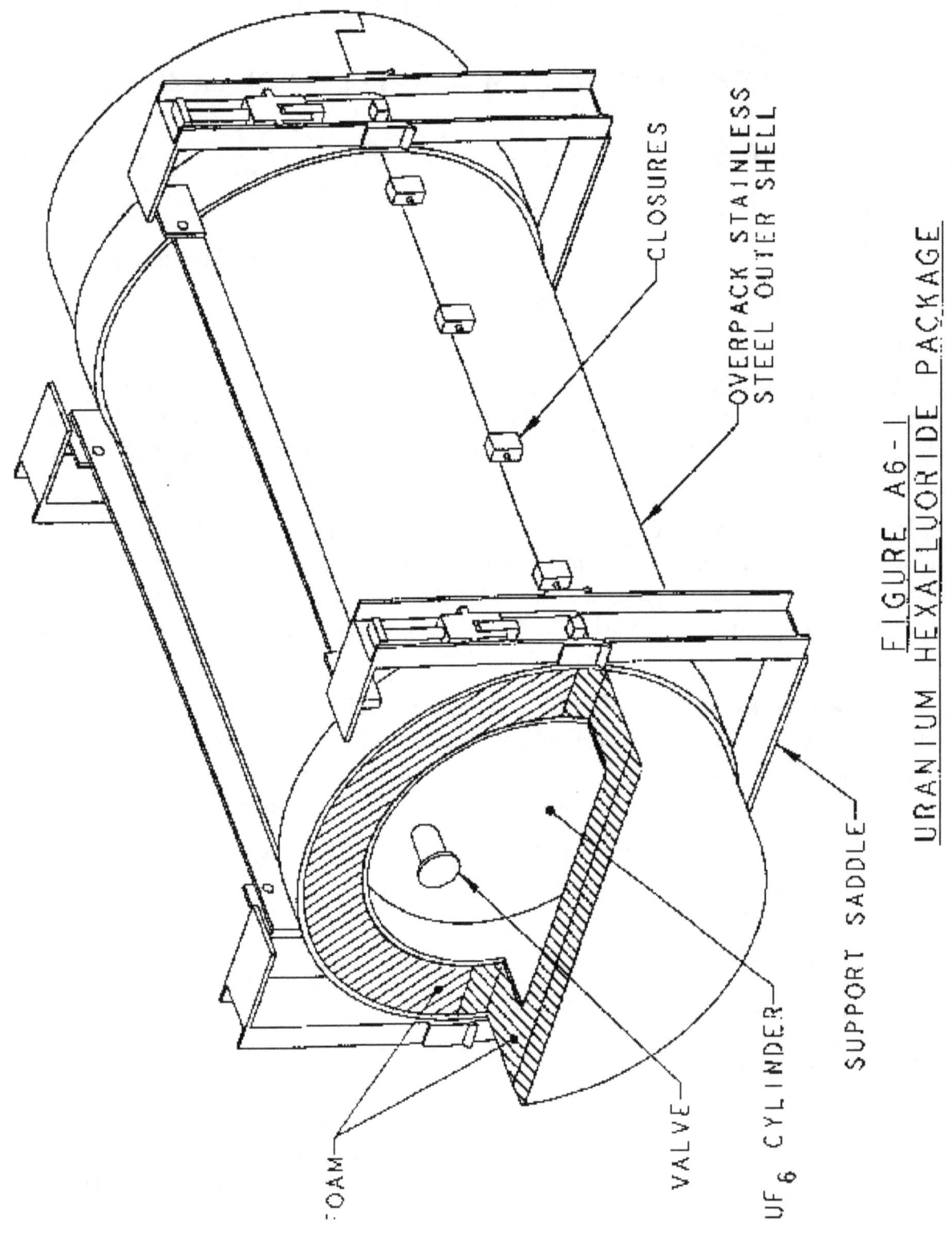

CLOSURES

OVERPACK STAINLESS
STEEL OUTER SHELL

SUPPORT SADDLE

VALVE

UF$_6$ CYLINDER

FOAM

FIGURE A6-1
URANIUM HEXAFLUORIDE PACKAGE

The cylinder is defined as the containment boundary for the UF_6. Unirradiated uranium enriched to less than 5% is a Type A quantity. Recycled uranium can be a Type B quantity due the presence of ^{232}U, ^{234}U, ^{236}U, and various radioactive impurities.

Shielding requirements are generally not significant because of the low radioactivity and self-shielding of UF_6. Compliance with regulatory limits for radiation levels is verified prior to shipment.

The overpack provides thermal protection to prevent overheating of the UF_6, which can cause hydraulic failure of the cylinder. The overpack also provides impact protection for the cylinder and the valve.

A6.2.2 Safety Features

- The steel cylinder precludes inleakage of water and provides containment under normal conditions of transport and hypothetical accident conditions.
- The cylinder skirt provides some protection to the valve during handling operations, normal conditions of transport, and hypothetical accident conditions.
- The overpack provides structural and thermal protection for the cylinder and its valve under hypothetical accident conditions.

A6.2.3 Typical Areas of Review for Package (Overpack) Drawings

- Overpack shell
  - Materials of construction
  - Dimensions and tolerances
  - Vents for pressure-relief of foam combustion products.
- Foam specifications
  - Type
  - Density
  - Compressive strength
  - Fire retardant characteristics
  - Limit on free chlorides.
- Closure devices
  - Torque
  - Valve protection device.

A6.2.4 Typical Areas of Safety Review

- The structural review concentrates on the ability of the overpack to protect the valve under hypothetical accident conditions.

- The structural and thermal reviews address the ability of the overpack to provide protection to the cylinder itself under hypothetical accident conditions. Because of the heat capacity of the UF_6, a partially filled cylinder may be more susceptible to hydraulic failure than a full cylinder.

- The containment review verifies that the cylinder meets the containment criteria in ANSI N14.5 for Type B packages.

- The criticality review confirms that there is no water inleakage under normal conditions of transport and hypothetical accident conditions. The minimum transport index for criticality control is specified in 49 CFR 173.417.

- The review of operating procedures ensures that the valve is properly closed and leak tested, as appropriate, and that the valve protection device, if applicable, is installed. This review also confirms that the radiation levels are verified to meet the regulatory limits prior to transport.

- The review of the acceptance tests and maintenance program evaluates the inspection procedures for the overpack, including the physical condition of the inner and outer shells, corrosion, performance of the foam during the service life of the overpack, and wear of cushioning pads between the cylinder and overpack. The review also verifies that the cylinder is tested and maintained in accordance with the requirements in 49 CFR 173.420 and ANSI N14.1.

References

American National Standards Institute, ANSI N14.5-1997, "American National Standard for Radioactive Materials–Leakage Tests on Packages for Shipment," New York.

Institute of Nuclear Materials Management, "American National Standard for Nuclear Materials—Uranium Hexafluoride–Packaging for Transport," ANSI N14.1-1995, New York.

U. S. Enrichment Corporation, "Uranium Hexafluoride: A Manual of Good Handling Practices," USEC-651 (Revision 7), January 1995.

APPENDIX A7:

HIGH ENRICHED URANIUM OR PLUTONIUM PACKAGES

A7.1 Package Type

A7.1.1 Purpose of Package

The purpose of this type of package is to transport Type B quantities of high enriched uranium or plutonium (other than by air).

A7.1.2 Description of a Typical Package

A typical packaging consists of one or two containment vessels and an outer container. Double containment is required for plutonium in excess of 20 Ci, except as specified in §§71.63(b)(1-3).

The outer container is a steel drum with a removable head and weather-tight gasket. The head is usually secured by a clamp ring with a tamperproof seal. Vent holes near the top of the drum, which provide pressure relief under hypothetical accident conditions, are capped or taped during normal conditions of transport to prevent water inleakage.

The inner containment vessel is a steel container, typically a stainless steel cylinder, with a maximum outer diameter of 0.127 m (5 in.), closed by a welded bottom cap and a welded top flange with a bolted lid. The lid, which is sealed by two O-rings, contains a leak-test port and sometimes a separate fill port for leak testing. Unless double containment is provided, this containment vessel is centered in position inside the outer container by fiberboard (or similar material) insulating material. If double containment is required, the inner (primary) containment vessel is positioned inside a secondary containment vessel.

The contents are uranium or plutonium, typically in metal, oxide, or nitrate form. The uranium or plutonium is generally placed in plastic bags or metal cans prior to loading into the containment vessel. Spacers are often used to maintain the position of the contents. Uranium may be in liquid form. Plutonium in excess of 20 Ci must be shipped as a solid.

A sketch of a typical package for high enriched uranium is presented in Figure A7-1. A package for plutonium would be similar, except than a second containment system would be required.

A7.2 Package Safety

A7.2.1 Safety Functions

The principal functions of the package are to provide containment and criticality control.

Package design features that accomplish the containment and criticality functions generally also provide adequate shielding to satisfy the requirements for nonexclusive-use shipment. Additional shielding may be required if significant quantities of certain isotopes, e.g., ^{238}Pu or ^{241}Am (from the decay of ^{241}Pu) are present.

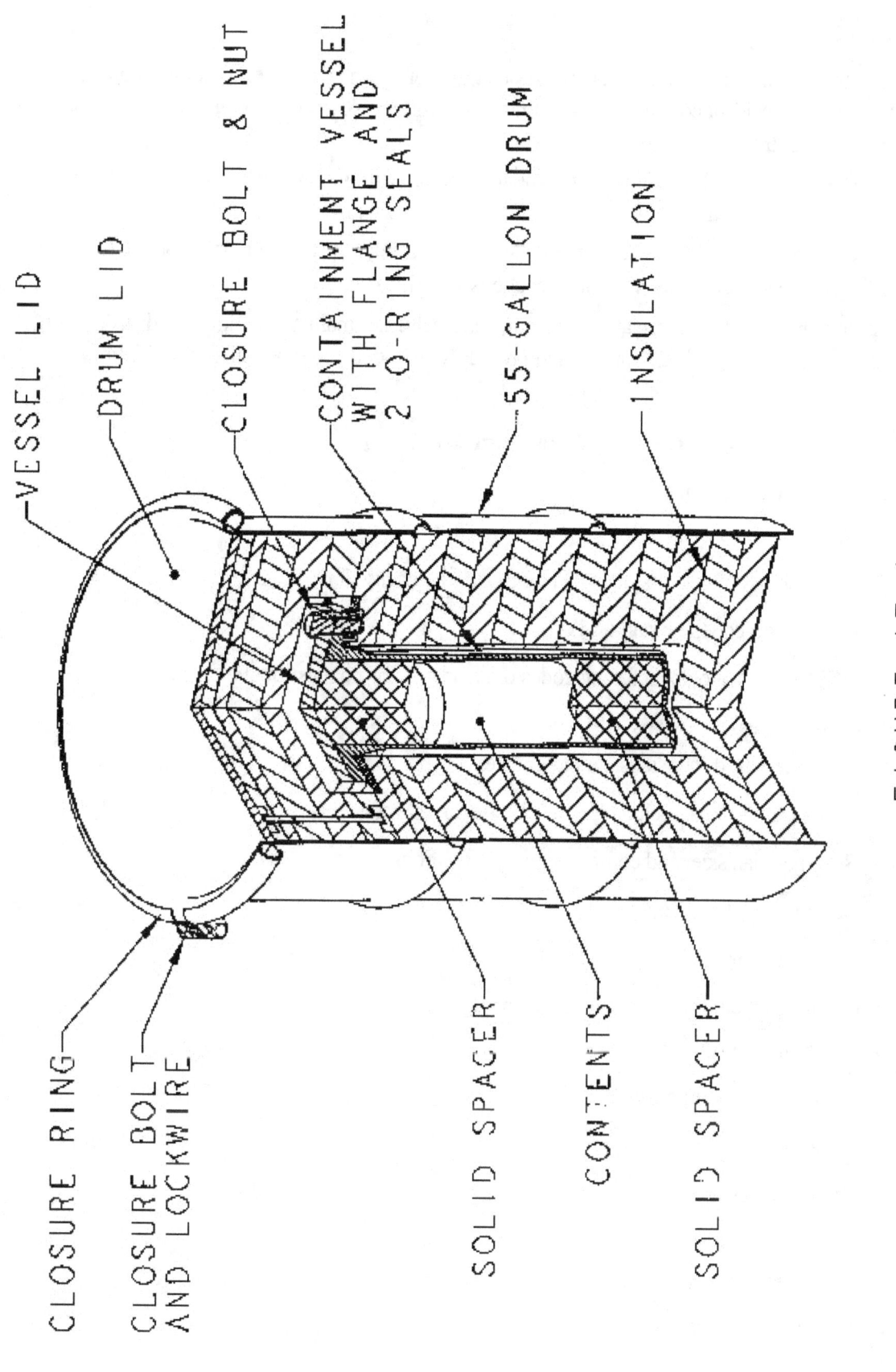

FIGURE A7-1

HIGH ENRICHED URANIUM PACKAGE

VESSEL LID

DRUM LID

CLOSURE BOLT & NUT

CONTAINMENT VESSEL WITH FLANGE AND 2 O-RING SEALS

55-GALLON DRUM

INSULATION

CLOSURE RING

CLOSURE BOLT AND LOCKWIRE

SOLID SPACER

CONTENTS

SOLID SPACER

A7.2.2 Safety Features

- The steel drum and insulating material protect the containment vessel and contents under hypothetical accident conditions and maintain a minimum spacing between packagings for criticality control.

- The steel inner vessel provides containment of the radioactive material. An additional containment vessel also provides containment for plutonium, if required by §71.63(b).

- The diameter and volume of the inner containment vessel, together with limits on the fissile mass of the contents, ensure that a single package is subcritical.

- The containment vessel, insulating material, and steel drum maintain a minimum distance from the contents to the package surface and provide some attenuation to satisfy the shielding requirements.

A7.2.3 Typical Areas of Review for Package Drawings

- Containment vessel body
  - Materials of construction
  - Dimensions and tolerances, including maximum cavity dimensions
  - Fabrication codes or standards
  - Weld specifications, including codes or standards for nondestructive examination.

- Containment vessel closures
  - Lid materials, dimensions, and tolerances
  - Bolt specifications, including number, size, and torque
  - Seal material, size, and compression specifications
  - Seal groove dimensions
  - Leak-test ports.

- Spacers to position or displace fissile material
  - Material of construction
  - Dimensions and tolerances
  - Locations.

- Insulating material
  - Type
  - Dimensions and tolerances
  - Density.

- Outer drum
  - Material
  - Closure, including use of heavy-duty clamp ring, bolt torque
  - Dimensions
  - Applicable codes or standards.

A7.2.4 Typical Areas of Safety Review

- The structural review confirms that packaging integrity is maintained under the drop, crush, and puncture tests. The review also verifies that the drum lid remains securely in place.

- The structural and thermal reviews evaluate the performance of the containment system under both normal conditions of transport and hypothetical accident conditions. Primary emphasis is on the structural integrity of the inner vessel and its closure, and on the thermal performance of the O-rings. If the package provides double containment, each containment vessel must separately meet the containment criteria.

- The structural and thermal reviews address the condition of the package and the minimum spacing between different packages under hypothetical accident conditions. Damage to the outer drum and charring of the insulating material may result in closer spacing than that of normal conditions of transport.

- The thermal and containment reviews verify that the hydrogen concentration in any confined volume will not exceed 5% (by volume) during a period of one year. Shorter time periods have been approved based on detailed operating procedures to control and track the shipment of packages.

- The criticality review addresses in detail both normal conditions of transport and hypothetical accident conditions. Key parameters for this review include the number of packages in the arrays, array configuration (pitch, orientation of packages, etc.), positioning of the containment vessels within the drum, moderation due to inleakage of water, the condition and quantity of spacing material, and interspersed moderation between packages.

- The review of operating procedures confirms that the containment vessels have been properly closed and bolts torqued, and that an appropriate pre-shipment leak test is performed.

- The review of the acceptance tests and maintenance program verifies that appropriate fabrication and periodic verification leakage tests are performed.

APPENDIX A8:

TYPE B SPECIAL FORM PACKAGES

A8.1 Package Type

A8.1.1 Purpose of Package

The purpose of this type of package is to transport a Type B quantity of radioactive material in special form.

A8.1.2 Description of a Typical Package

A typical packaging consists of a cask body with a lid, base, and protective jacket.

The cask body is a lead-filled cylinder with a stainless steel inner and outer shell. A drain tube penetrates the cavity and is sealed with a plug, which is covered by the protective jacket during transport. A lead-filled, stainless steel lid is bolted to the tapered top of the main body and sealed with a weather-tight gasket. Both the body and the lid generally have lifting devices that are covered during shipment by the protective jacket.

The base is a square steel skid that bolts to the protective jacket. The skid consists of energy-absorbing steel angles (stiffeners). Several I-beams are welded to the base to enable handling by a forklift.

The protective jacket is a double-walled steel cylinder with an open bottom and a protruding box section positioned diametrically across the top and vertically down the sides. The jacket may contain thermal insulation. A steel flange bolts to the base, and the main body of the packaging is centered within the jacket by steel tubes welded to the jacket inner wall. Steel lifting loops are typically welded to the top corners, and tie-down devices are welded to the sides.

The contents of the package typically consists of byproduct material in special form.

A sketch of a typical Type B special form package is presented in Figure A8-1.

A8.2 Package Safety

A8.2.1 Safety Functions

The principal safety function of the package is to provide radiation shielding. Containment is provided primarily by the special form source itself. The packaging must maintain the sources in the fully shielded configuration under normal conditions of transport and hypothetical accident conditions.

A8.2.2 Safety Features

- The lead shield provides shielding for gamma radiation.
- The protective jacket provides structural and thermal protection to the main body, which contains the special form radioactive material.

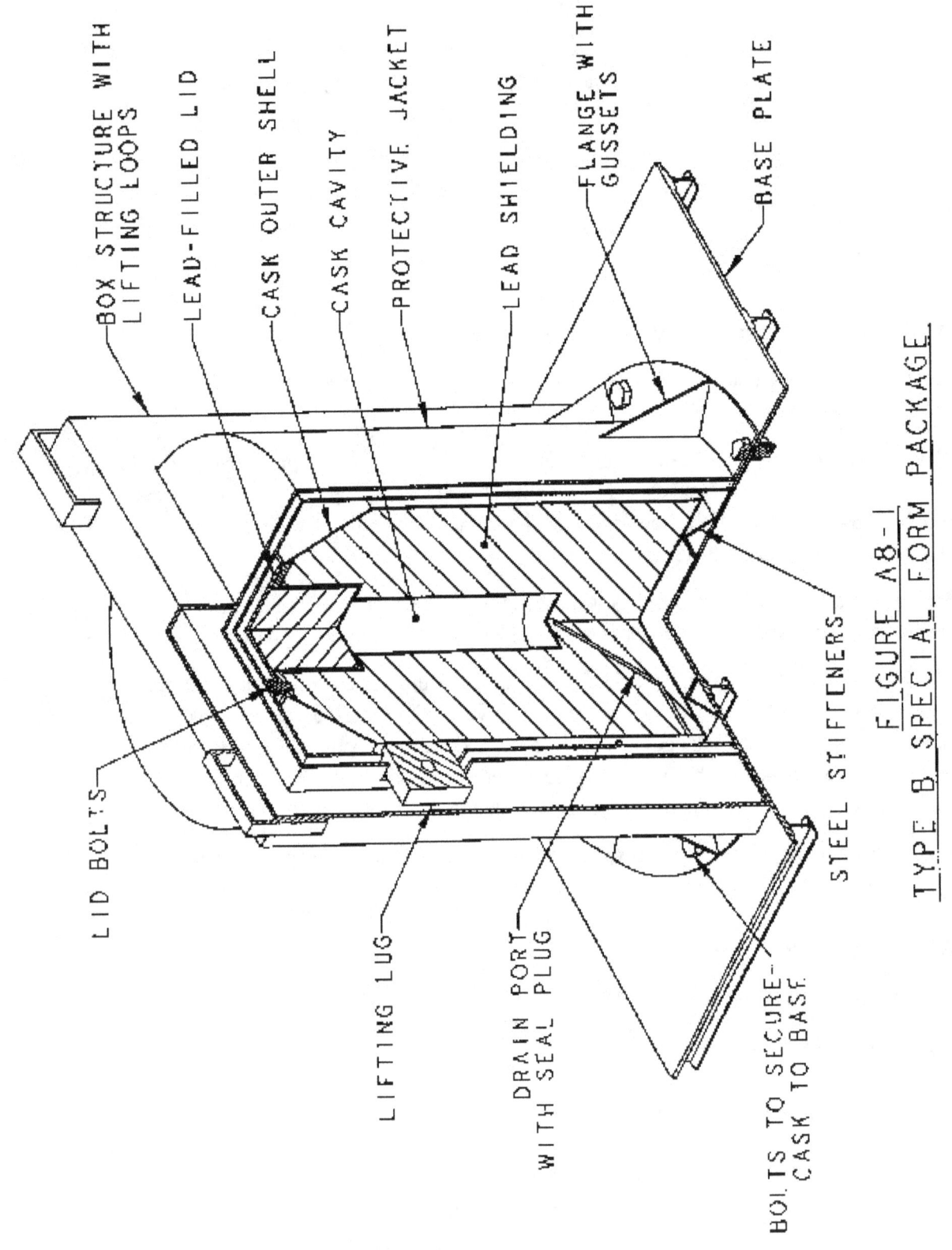

BOX STRUCTURE WITH LIFTING LOOPS

LEAD-FILLED LID

CASK OUTER SHELL

CASK CAVITY

PROTECTIVE JACKET

LEAD SHIELDING

FLANGE WITH GUSSETS

BASE PLATE

LID BOLTS

LIFTING LUG

DRAIN PORT WITH SEAL PLUG

BOLTS TO SECURE CASK TO BASE

STEEL STIFFENERS

FIGURE A8-1

TYPE B SPECIAL FORM PACKAGE

A8.2.3   Typical Areas of Review for Package Drawings

- Cask body
  - Materials of construction
  - Dimensions and tolerances of steel shells and lead shield
  - Fabrication codes or standards, including any special processes for lead pour
  - Weld specifications, including codes or standards for nondestructive examination.

- Closures
  - Lid materials, and their dimensions and tolerances
  - Bolt specifications, including number, size, minimum thread engagement, and torque
  - Seal material, size, and compression specifications
  - Seal groove dimensions
  - Vent and leak-test ports, including closure methods.

- Protective jacket
  - Method of attachment
  - Bolt specifications, including number, size, minimum thread engagement, and torque
  - Insulating material.

A8.2.4   Typical Areas of Safety Review

- The review of the general information verifies that the contents are special form.
- The structural and thermal reviews evaluate the ability of the shield to perform its intended function under normal conditions of transport and hypothetical accident conditions. Lead slumping should be inconsequential and the lead should not melt. These reviews ensure that the package has been tested under the most damaging conditions (e.g., impact orientation). The integrity of the cask closure and bolts is also reviewed.
- The thermal review should verify that no credit has been taken for the presence of helium in gaps between packaging components. The review should verify that the heat transfer medium is air, and that the effects of air on the contents and packaging components have been addressed.
- The shielding review evaluates the ability of the package to satisfy the allowed radiation levels during both normal conditions of transport and hypothetical accident conditions.
- The review of operating procedures verifies that the cask has been appropriately drained and that the bolts are properly torqued.
- The review of the acceptance tests and maintenance program ensures that appropriate tests are specified for shielding and thermal performance.